This beautiful book will arm you with everything you need to incorporate the abstract into your own work, through experimentation and process.

The first chapter explores inspiration from nature, always a rich resource for sampling and developing new work. This chapter features the work of artists who are inspired by the natural world, including Jan Beaney and Jean Littlejohn. Chapter two explores shapes, colour and mid-century influences, including projects such as how to make a patchwork bag and rubber stamp prints. Chapters three and four look at surface pattern, mixed media and line art, featuring sections on kantha and boro techniques, stencil and screen printing and making a 'junk' collage. Chapter five explores how to capture meaning in your practice, while the final chapter looks at playing and process, to help you to make and experience your own personal responses to the theme.

Featuring the work of numerous leading textile artists from all over the world, including Dail Behennah, Michelle House, Jessie Cutts, Sarah Z. Short and Iain Perry, this fascinating and evocative book is suitable for all ages and abilities, from students to experienced practitioners, or anybody seeking to inject new inspiration into their textile practice.

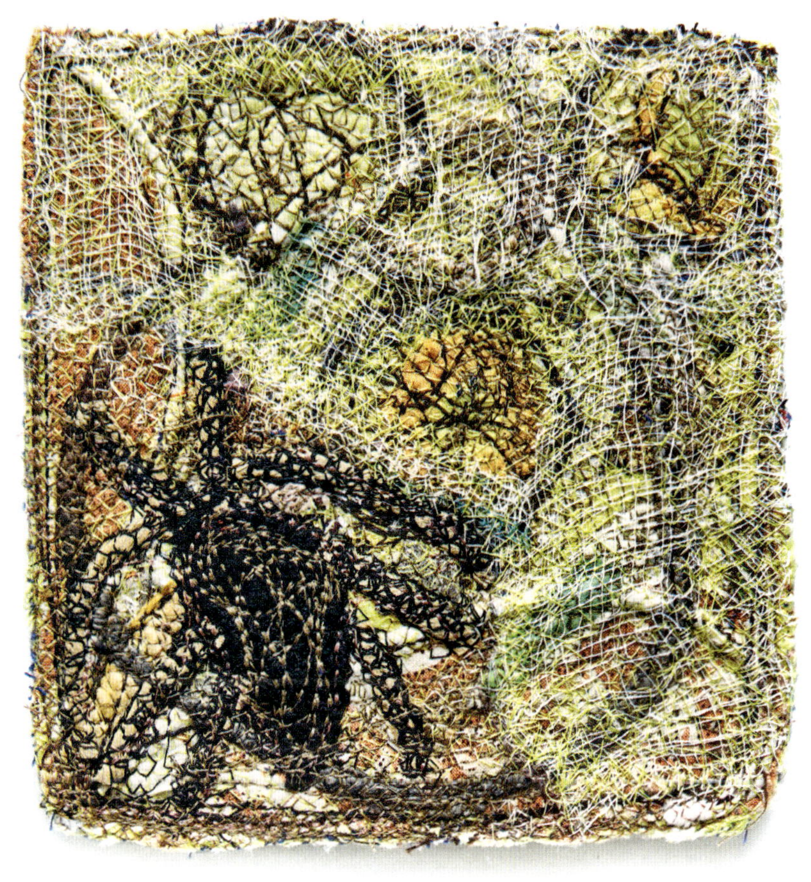

ABSTRACT TEXTILES

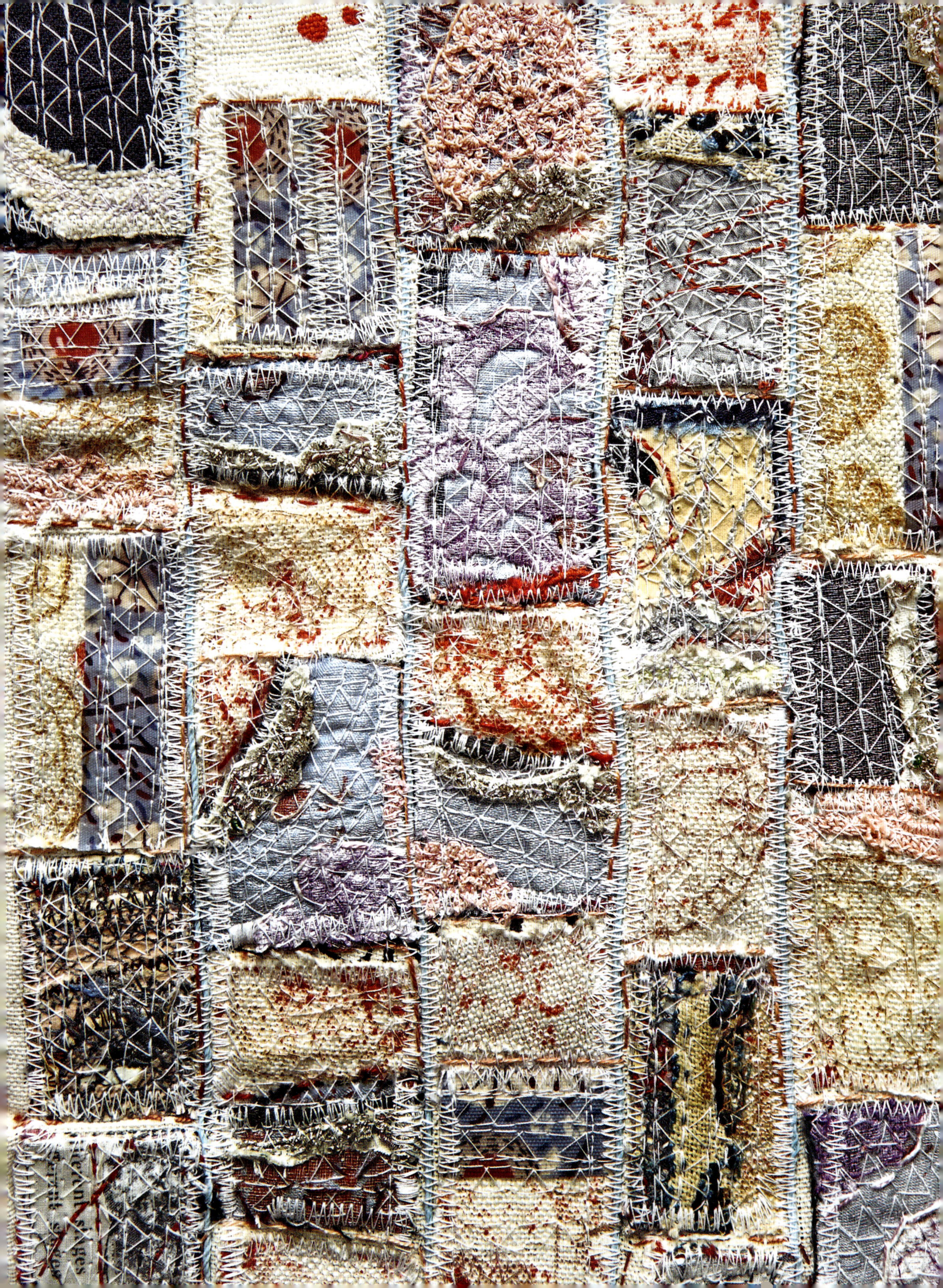

ABSTRACT TEXTILES

COLOUR, SHAPE AND PATTERN IN TEXTILE ART

ANNE KELLY

BATSFORD

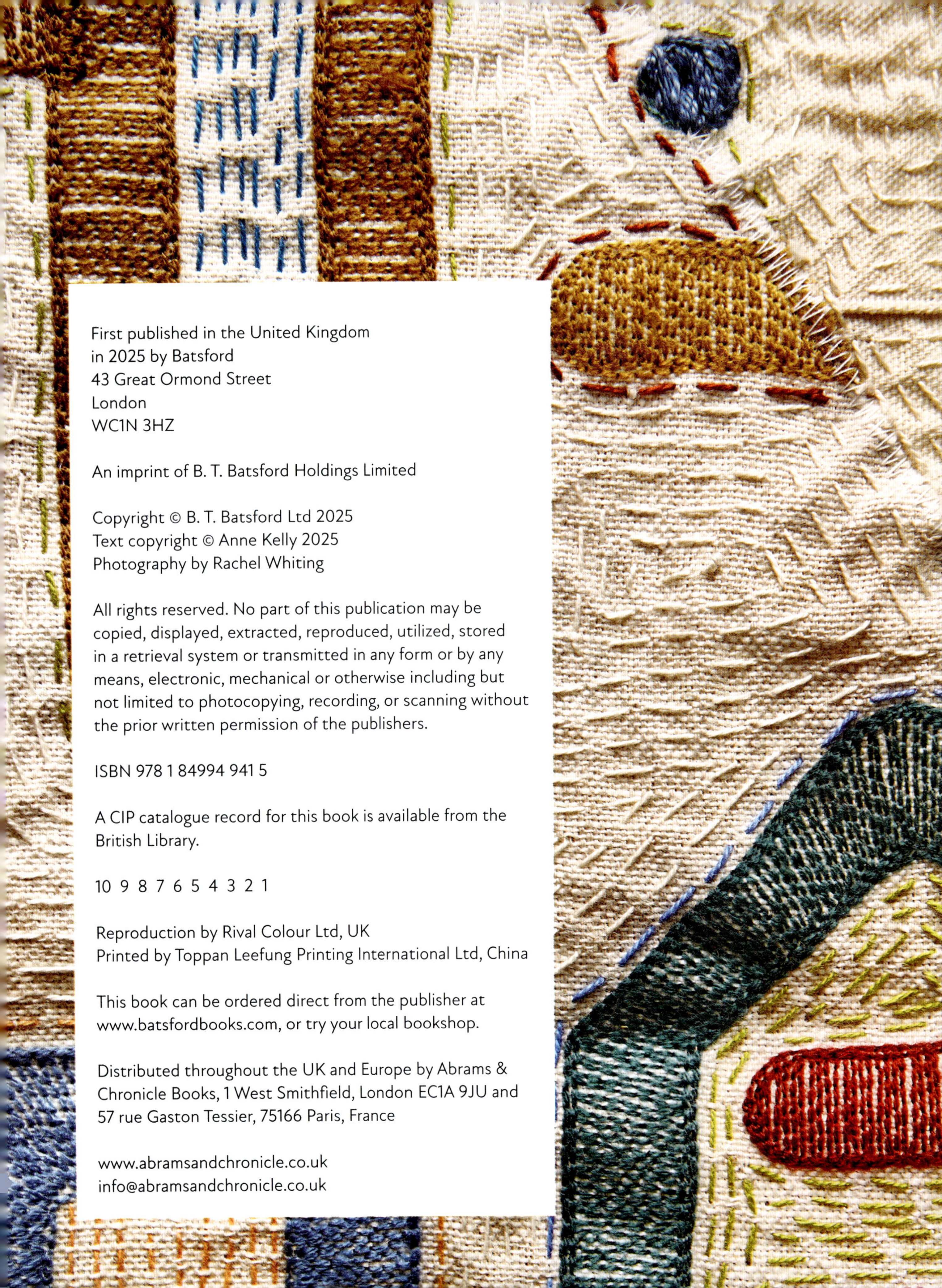

First published in the United Kingdom
in 2025 by Batsford
43 Great Ormond Street
London
WC1N 3HZ

An imprint of B. T. Batsford Holdings Limited

Photography by Rachel Whiting

ISBN 978 1 84994 941 5

A CIP catalogue record for this book is available from the British Library.

10 9 8 7 6 5 4 3 2 1

Reproduction by Rival Colour Ltd, UK
Printed by Toppan Leefung Printing International Ltd, China

This book can be ordered direct from the publisher at www.batsfordbooks.com, or try your local bookshop.

Distributed throughout the UK and Europe by Abrams & Chronicle Books, 1 West Smithfield, London EC1A 9JU and 57 rue Gaston Tessier, 75166 Paris, France

www.abramsandchronicle.co.uk
info@abramsandchronicle.co.uk

CONTENTS

Opposite: Anne Kelly, *Kantha Sample* (detail). Stitched textile

'With the most primitive means the artist creates something which the most ingenious and efficient technology will never be able to create.'

Kazimir Malevich (1879–1935)

INTRODUCTION

Abstract textile work is a popular and enduring theme in textile art and can be viewed in many ways. From Amish quilts through to contemporary makers, this genre is captivating because of colour and shape, and its ability to capture mood and feeling. In this book, we will show you how to progress from the realistic to the abstract through experimentation and process. From shapes, colours, surface pattern, line and art/textile influences, you will be able to experience and convey your own personal responses to the theme.

As in my previous publications, I combine examples of my own work and projects with examples from exciting contemporary artists and makers. This book is suitable for practitioners of all ages and abilities.

In **Chapter One: From Nature to Abstraction** we start from a representational viewpoint and use familiar shapes found in the natural world to create mixed media textile collages, looking at the work of artists and makers who use nature as their starting point.

Chapter Two: Shapes and Colour – Mid-century Influences starts with the building blocks of shape and colour to create original designs that can be applied to a variety of surfaces. The influence of pivotal women artists from the twentieth century informs and endures in current design work.

Chapter Three: Surface Pattern and Mixed Media features an exciting collection of mixed media works, inspired by shapes and materials. This chapter also looks at surface pattern and printing, and the combination of random elements to create original and creative designs.

In **Chapter Four: Following the Line** we take a considered approach to linear compositions, both in design and stitch terms. Line is another building block that enables structure and composition to develop organically in abstract work.

Chapter Five: Capturing Meaning discusses the challenging issue of how to inject emotion into abstract work. Through exploration and repetitive, intentional mark-making, we are able to create personal and individual responses and work through themes.

Chapter Six: Playing and Process shows examples of artists who use their practice to explore media through creative play. This is a 'workbook' chapter and should provide helpful reference for practitioners starting out on their abstract journey.

Above: Anne Kelly, *Window*, 1982. Black and white photograph

Opposite: Anne Kelly, *Mid-century Meadow Jacket* (detail). Stitched and appliquéd jacket

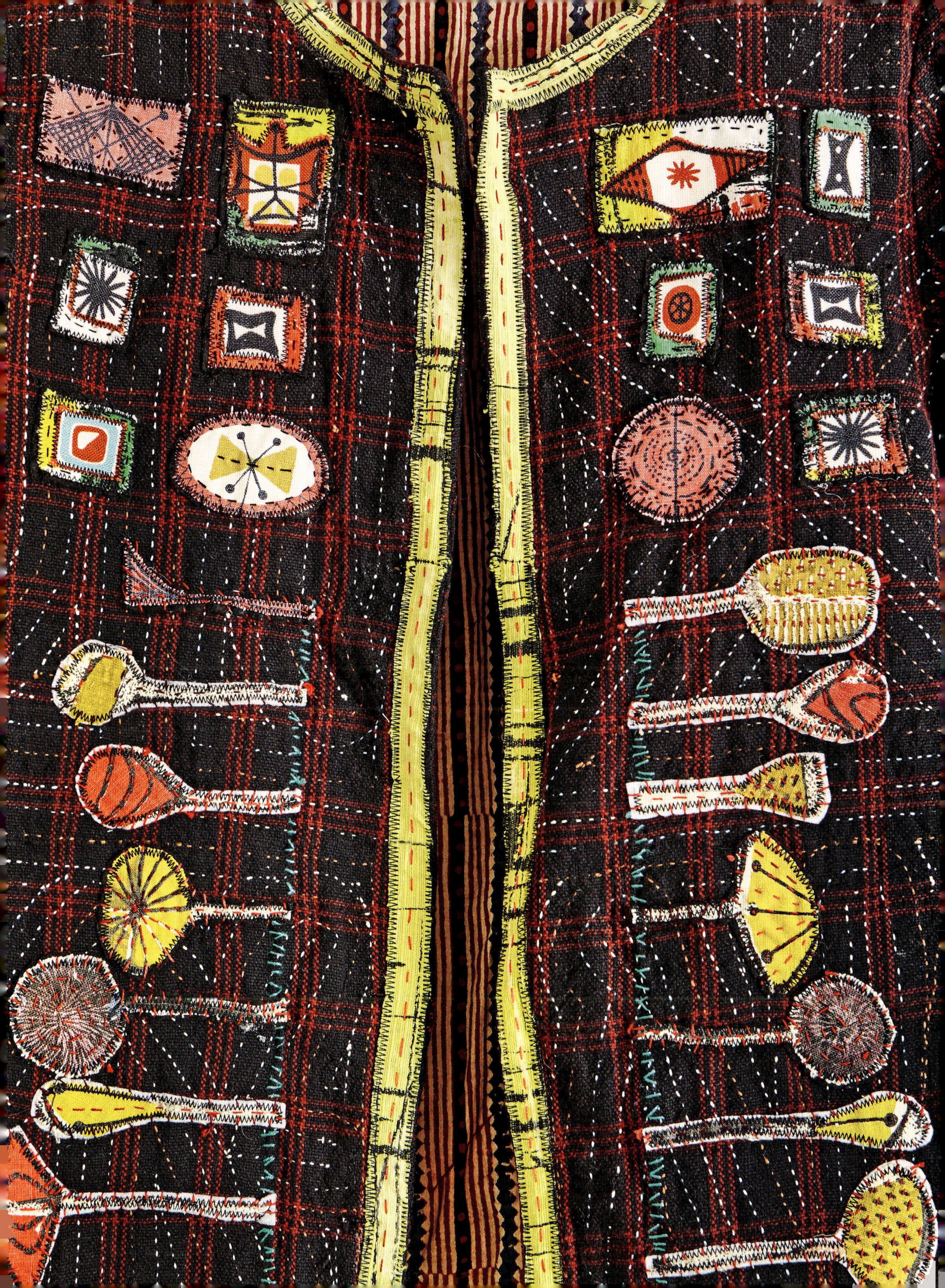

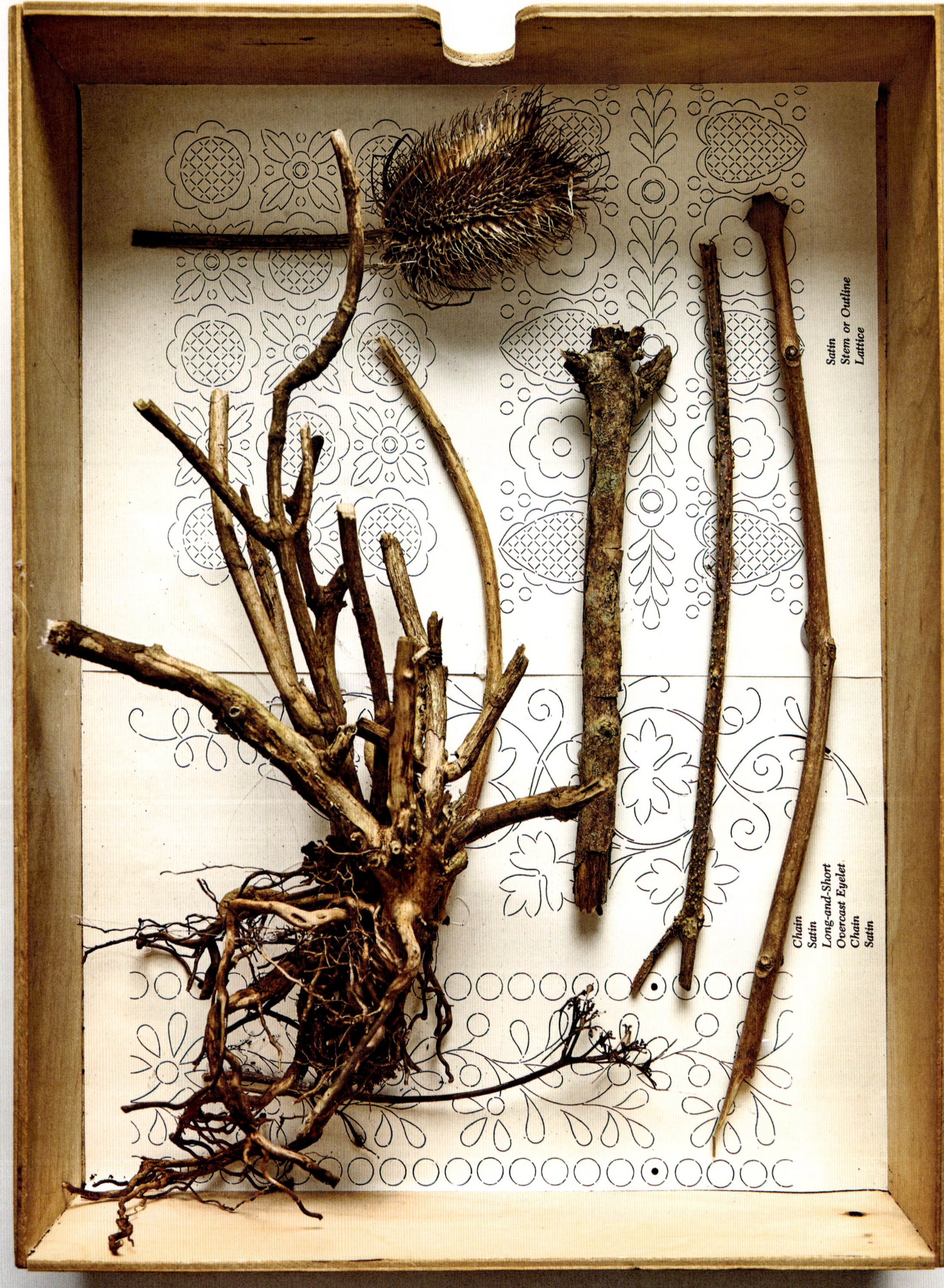
Satin
Stem or Outline
Lattice
Chain
Satin
Long-and-Short
Overcast Eyelet
Chain
Satin

CHAPTER 1

FROM NATURE TO ABSTRACTION

'For me nature is not landscape, but the dynamism of visual forces... (which) can only be tackled by treating colour and form as ultimate identities, freeing them from all descriptive or functional roles.'

Bridget Riley (b.1931)

Opposite: Anne Kelly, *Inspiration Box, Natural Forms*. Mixed media in wooden box

INSPIRATION FROM NATURE

Probably the most challenging element of approaching abstraction is to know where to start. With representational art, there are natural starting points and a readymade vocabulary embedded in the history of art and making. With abstraction, there are signposts and references that can help us to find inspiration, but it requires organisation and focus.

The natural world is always a rich resource. There are many aspects of nature that are great for sampling and developing new work: colour, texture, line, form and pattern. I like to start with drawings, whether from photographs or from life. I begin with small sketches and studies, fragments of postcards and images. I like to experiment with paint and paper before progressing onto cloth.

In this chapter we look at artists who are inspired by the natural world and landscape, showing their transition from representational to abstract. This is a good place to start and work through before attempting more complex work. Sketchbooks and notebooks are useful to make studies and develop ideas, and enable you to make a range of versions and decisions.

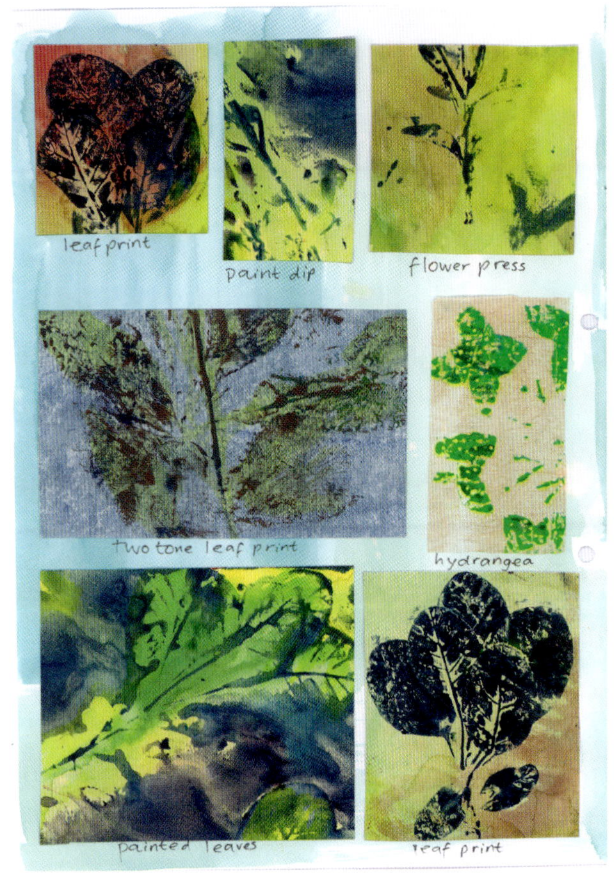

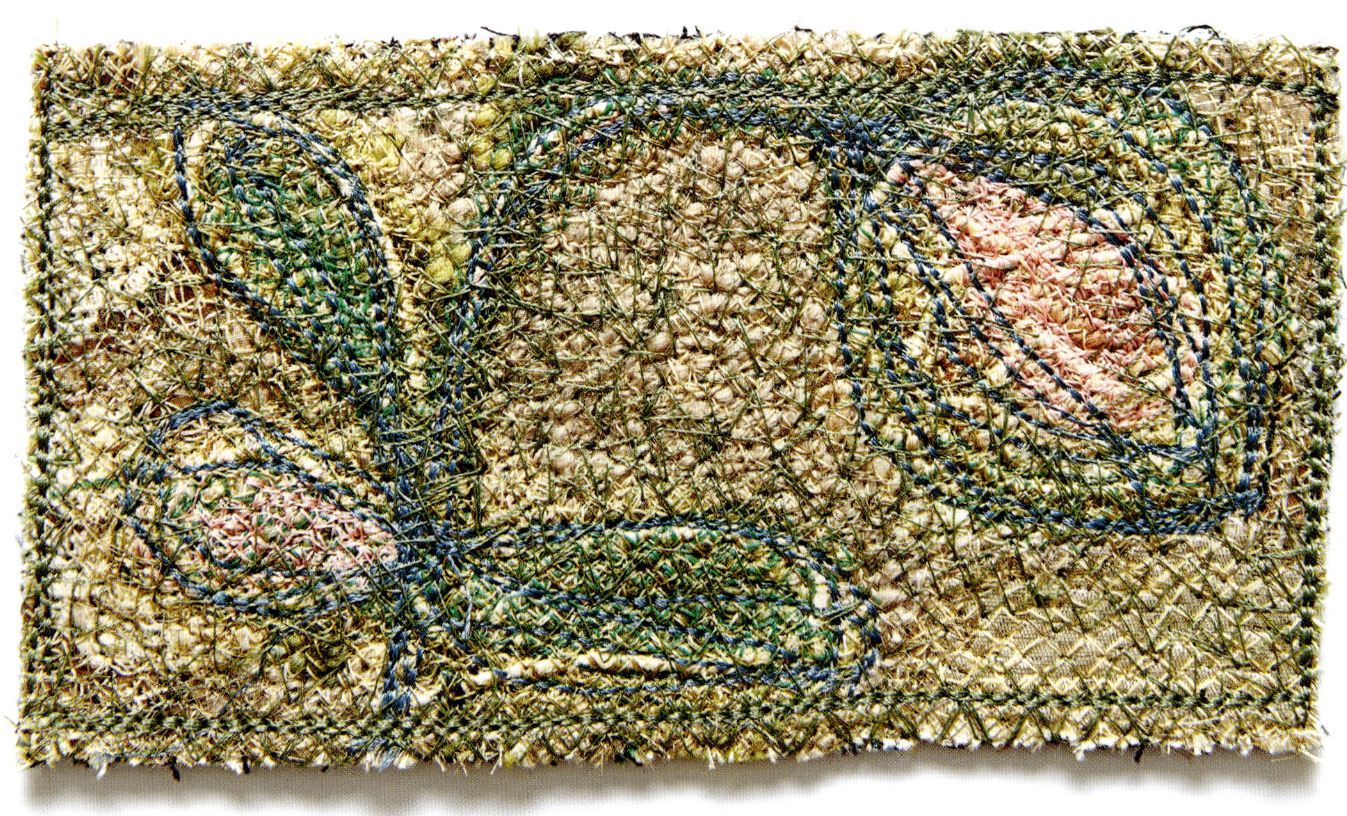

SAMPLING SKETCHBOOK PAGES

Here are some examples of experiments and starting points from my sketchbooks. To make similar studies, you will need:

- Simple natural forms, interesting shapes, plants, flowers
- Sketchbook or cartridge paper
- Pencil
- Ink pens (I used a waterproof fine-line pen)
- Watercolour or gouache paint, brushes
- Waste paper and tissue
- Glue stick
- Small shape prints and ink pad (optional)

Starting points in nature

1. **Isolating shapes** – choose interesting shapes, whether flowers, leaves or plants. Simplify the shape and repeat, using a fine-line pen when happy with the pencil version. Colour with pencil and paint, with watercolour washes over the top.
2. **Mini collage** – this is useful for starting to create compositions. Choose a colourway that goes with your natural form. Cover your background with pieces of torn paper. Draw your simplified forms over the top, then add colour.
3. **Simple blocks** – use the paint to create a wash over your background. When dry, use the simplified shapes of your plants to draw over the top, using a pencil then fine-line pen. Finally, use the shaped blocks to print over the top.

Opposite top: Anne Kelly, *Sketchbook Pages*. Mixed media

Opposite bottom: Anne Kelly, *Flower*. Mixed media embroidery

Below: Anne Kelly, *Sketchbook Pages*. Mixed media

1

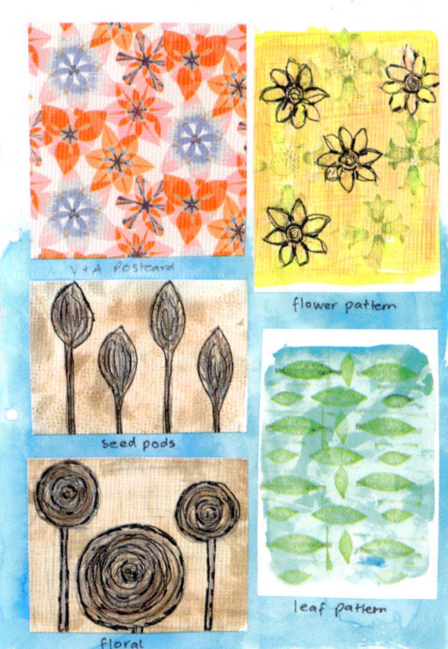

2

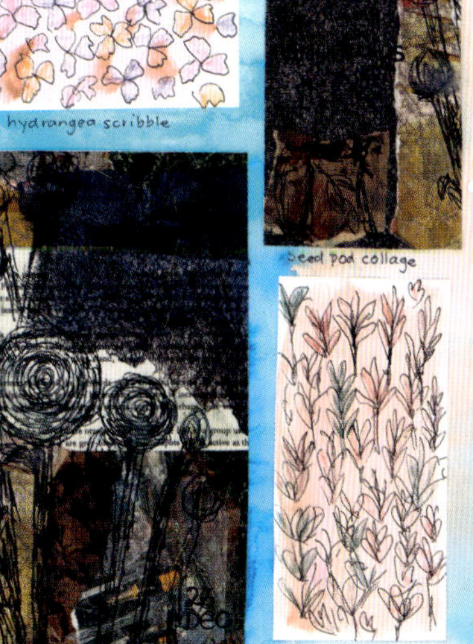

3

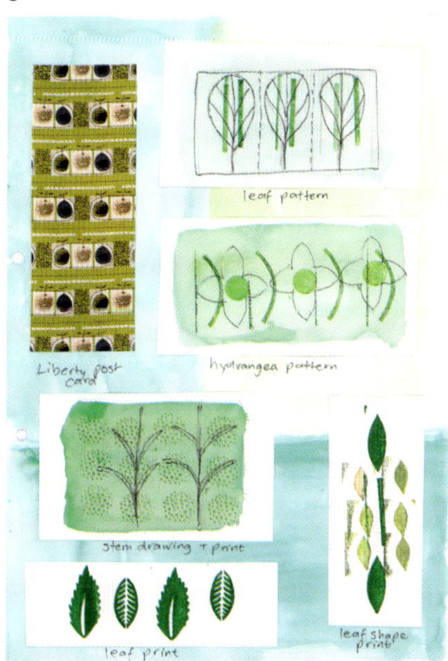

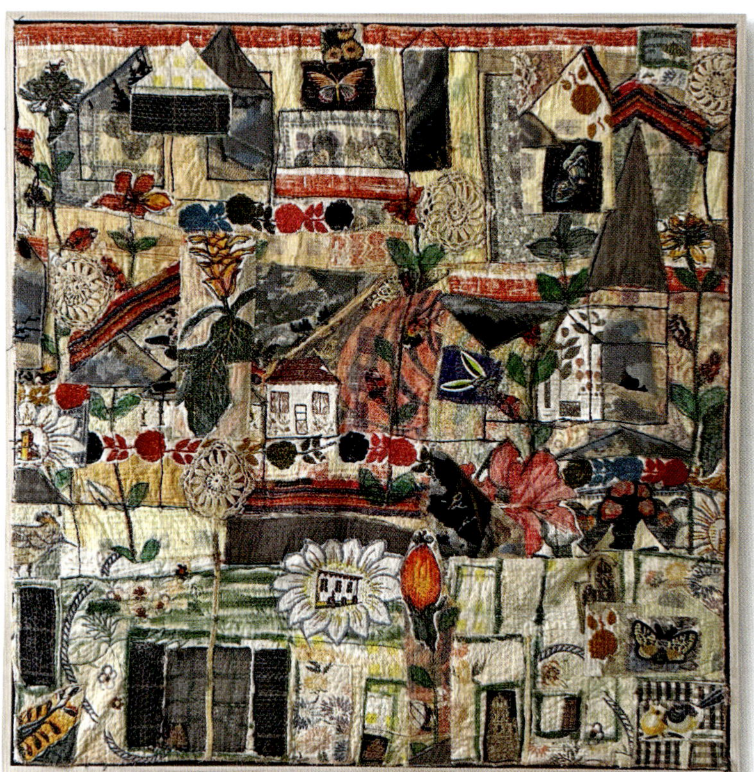

Left: Anne Kelly, *Italian Landscape*. Mixed media textile

FROM NATURE TO ABSTRACTION – AN ITALIAN LANDSCAPE

My large textile collage *Italian Landscape* is a piece based on sketchbook pages, made on a residency in Abruzzo, Italy. I decided to take elements from the sketches and develop them into shapes rather than transposing them as straight interpretations of the landscape. I used some previously worked pieces of textile to give the piece additional texture and colour. I also used pieces of fabric sourced in Italy.

Before I started making the piece, I sketched out the general outline shapes that I wanted to use. This helped to develop the scale of the piece and enabled me to cut out the pieces to the right size for collaging. I used a strong piece of canvas about 1m (40in) square. I then used the drawing to sketch out the various shapes and areas of the composition.

I pinned and tacked (basted) the elements of the collage onto the background. I was then able to stitch them on with a sewing machine using my signature overstitching technique, as well as some hand stitching to finish the piece. My overstitching technique uses a repetitive machine stitch (like an edging or finishing stitch) to cover and blend pieces together, usually with white thread. This has the effect of creating a 'filter' or blurring the layers so that they appear more homogenous. I use old sewing machines for this that can cope with joining several layers of fabric. If you want to try this technique, you can use a sample to see which patterns of stitch repeated together work for you.

Italian Landscape was then mounted onto a painter's canvas and framed. It has been exhibited in the UK and France.

There are recognisable elements in the piece, which act as guideposts to the subject matter. I would describe this as a transitional piece, a hybrid between representation and abstraction. I have included it in this chapter to share that journey.

LANDSCAPE

FEATURED ARTIST

Sarah Symes

Sarah Symes's work is inspired by the elemental beauty of the Sea to Sky region of British Columbia, Canada. This series explores the interplay and connection between land, sea and sky. She tells us more here:

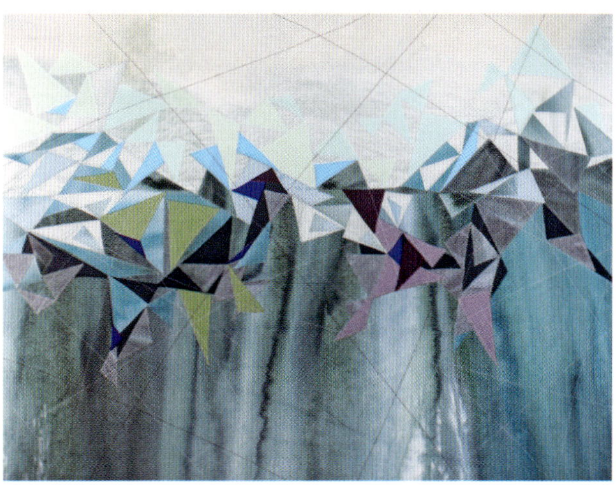

Above: Sarah Symes, *Sea to Sky Landscape No. 2.* Mixed media textile

Below: Sarah Symes, *Sea to Sky Landscape No. 4.* Mixed media textile

'Here are some of my sketchbook notes, recording my observations of the landscape: A dramatic cloudburst above Mount Garibaldi pulls the stormy sky into focus... The jagged rock faces of The Chief catch the evening sunlight, fracturing the line between rock and sky... A dense fog hovers above Howe Sound, obscuring the islands, leaving nothing between sea and sky... Heavy rains turn Shannon Falls into a thunderous waterfall, cascading in bubbly plumes blurring the shoreline... The alpine meadows of Sky Pilot are ablaze with wildflowers, leading to cool turquoise lakes high in the skyline... Snowmelt swells the Squamish River into an unstoppable torrent, flooding the white sandy beaches... Blue skies turn Lost Lake into a glittering mirror, reflecting snow-capped peaks against the sky...

'To create a texture representing the sky, I dyed fabric by soaking it in paint and leaving it outside in the rain. To create patterns representing water, I poured cups of water and dye over stretched fabric and let it drip dry. Each work of art started out with two pieces of background fabric, placed intentionally to section the picture plane with a horizontal line. I allowed the random dye markings to inspire the landscape and become part of the composition. I then cut a selection of smaller triangular pieces of fabric and appliquéd them to the background, working intuitively with silhouettes in mind to suggest recognisable landforms. This resulted in a geometric abstract formation, depicting the interplay and connection between landscape elements.

'I have titled the work numerically in order to track my progression through the idea. Across the body of work, I play with point of view by changing the canvas size and moving the horizon line. A low horizon gives the feeling of expansion and freedom and a high horizon gives the feeling of being surrounded by mountains. Rather than working from photos or sketches, I composed the pieces from memory, opening myself up to the inextricable link between the subject and my inner emotional landscape.'

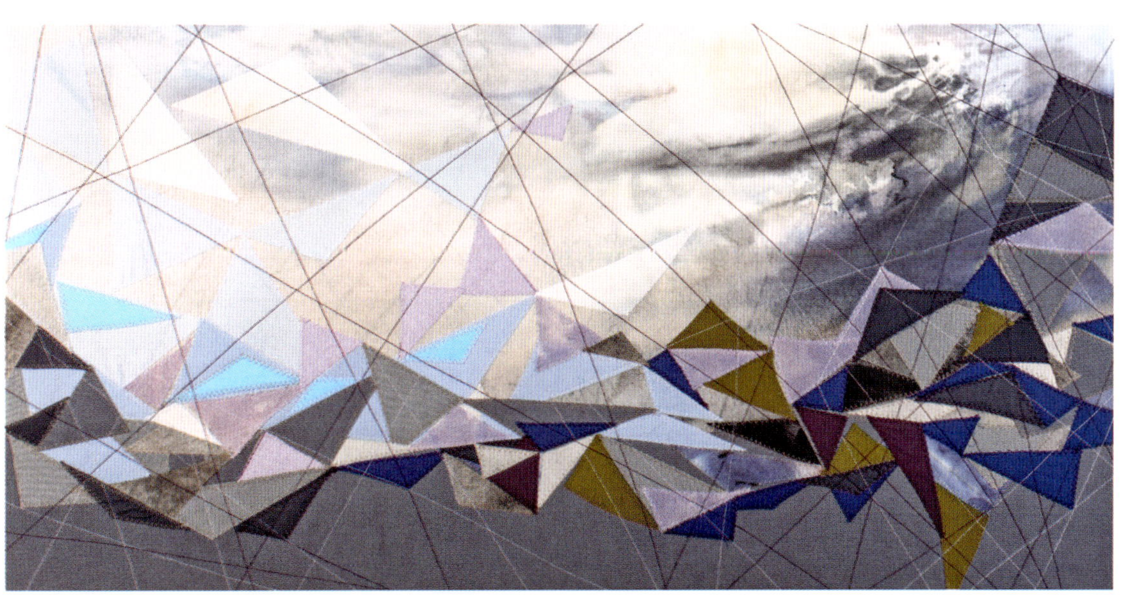

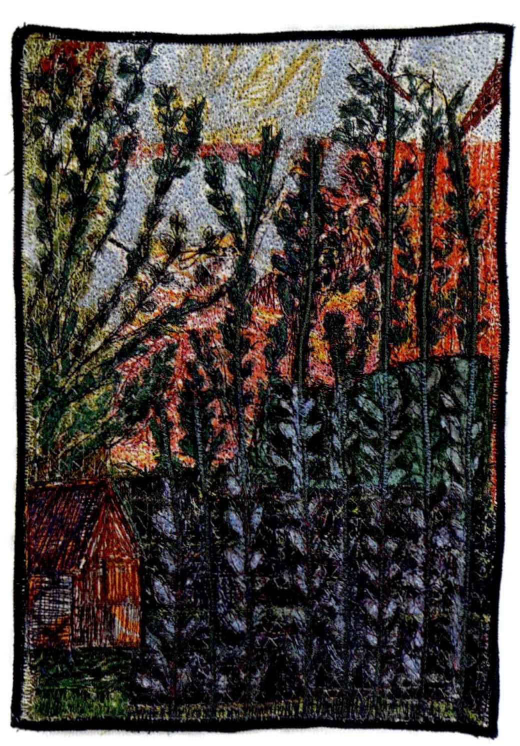

PHOTOS AS AN INSPIRATION AND RESOURCE

My piece *Suburban Gardens 2* was made when I was inspired by a photo of my garden at dusk in which the shades of the plants seemed to glow against an equally striking sunset, with beautiful colours. I simplified the shapes and colours with a sketch on a remnant of a woven cotton tea towel. I then painted the rough outlines onto the cotton with fabric paints. When it was dry, I drew over the patches of colour to create lines for stitching. I used free-motion embroidery with a dark grey thread to map out the main structure of the image. Then I used more colours in the background, matching my thread choices with the paints.

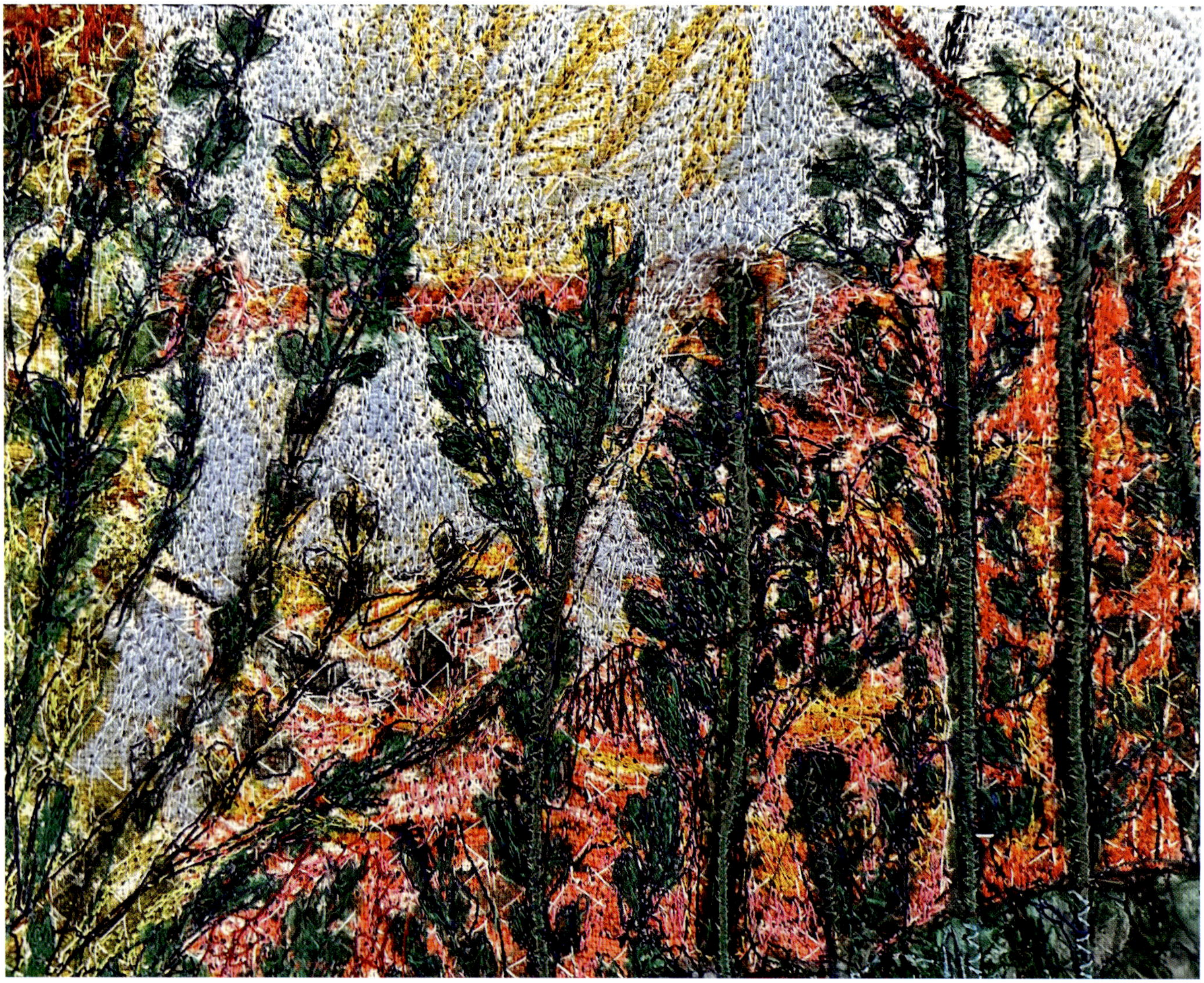

Left & below: Anne Kelly, *Sketchbook Pages*. Mixed media

Bottom: Anne Kelly, *Sketchbook Pages* (detail). Mixed media

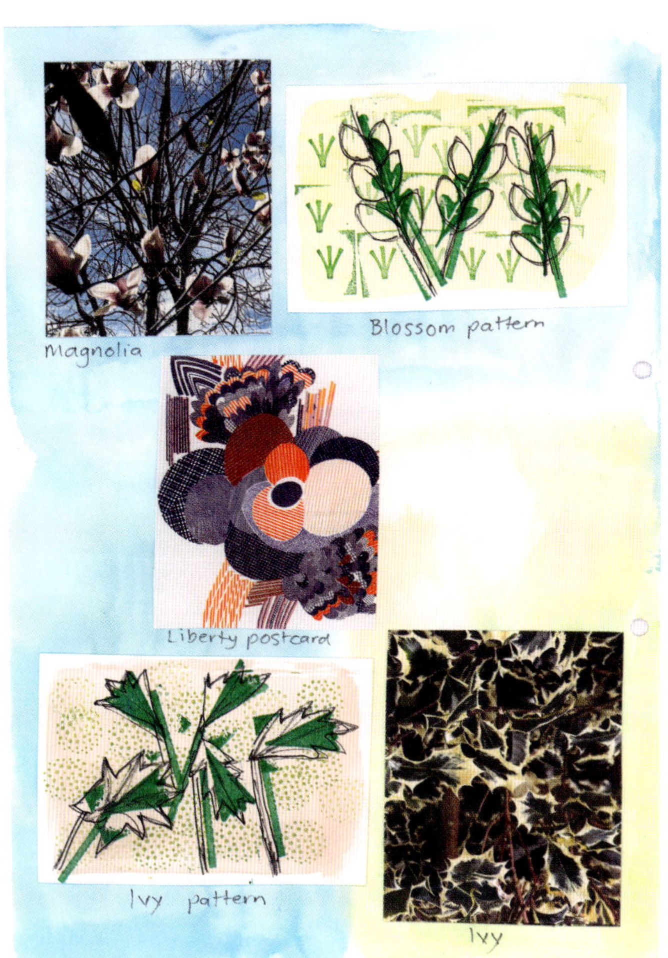

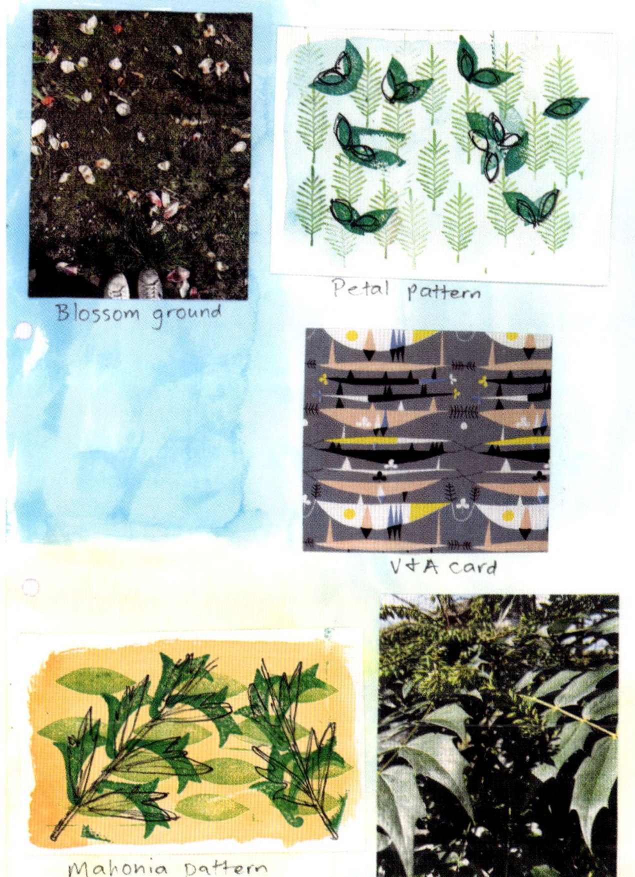

Opposite top: Anne Kelly, *Suburban Gardens 2*. Mixed media textile

Opposite bottom: Anne Kelly, *Suburban Gardens 2* (detail). Mixed media textile

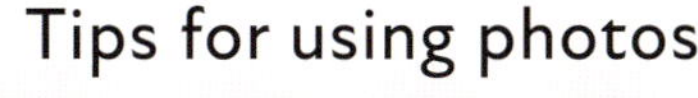

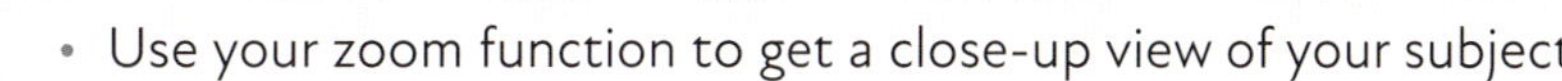

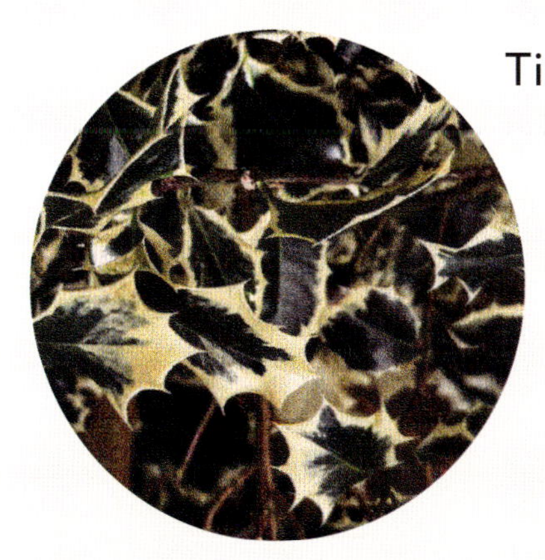

Tips for using photos

- Use your zoom function to get a close-up view of your subject.
- Print out your photos so you can work on them and from them.
- You can make collages and studies from your photos before you stitch.
- Work in stages, from the photo, to a drawing or collage, then a stitch sample.
- You can change the colours, shapes and orientation of your source image.

WORKING FROM THE LANDSCAPE

Jan Beaney and Jean Littlejohn have a venerable and well-established partnership, working, exhibiting and writing books about their practice, often inspired by the natural landscape. They are generous with suggestions and techniques, as we see here.

FEATURED ARTIST

Jan Beaney

'Our mantra is: the more you look, the more you see.

'Even if you do not want to draw, written descriptions and the odd diagram can also work well. So often a few words and some squiggles will remind you of the particular element you wish to remember. At times it is helpful to isolate your looking. For instance, look for stripes – perhaps rows of crops in a field, views through railings, patterns on a butterfly wing or creviced bark on a tree. Edges, spots or a colour could all be focused looking exercises. The list is endless.

Below: Jan Beaney, *Pale Waters*. Mixed media textile

Opposite top: Jan Beaney, *A Place of Joy, A Place of Sadness*. Mixed media textile

Opposite bottom: Jan Beaney, *Colour Study*. Mixed media textile

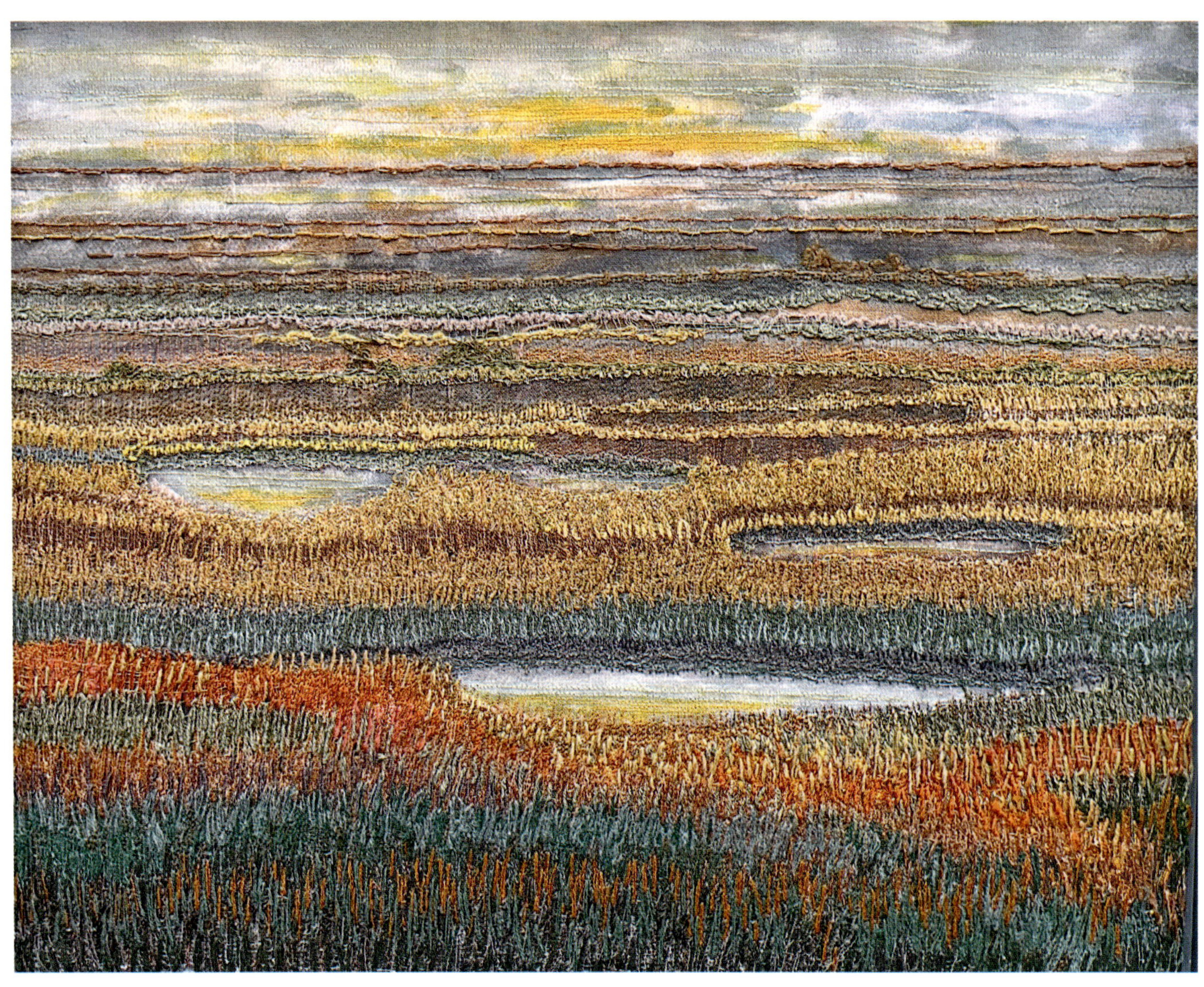

'Landscapes around the world have been inspirational as well as places just five minutes from home. Personally, colour is often the starting point, especially due to unusual light conditions. The question to ask is: what made you stop to have a longer look, take a photograph or get the urge to make a quick sketch? An unexpected colour scheme, a surprising textured surface, pattern or shape could be appealing.

'Rather than trying to copy every aspect or detail of the view, exaggerate the main image that attracted you in the first instance, and understate, suggest or even totally eliminate others. These actions will help you to develop your own style, capturing the essence of a particular location without being too literal.

'The works shown here have been created using several techniques. These include hand stitching on a dark cotton ground where broadly the image has been depicted with discharge paste or decolourants. Machine stitching was used to integrate all the surfaces. Other works have been created by hand and machine stitching on a variety of soluble materials to create a new fabric. Making a textured machine lace on a soluble material and applying it to a cotton ground has been another method appropriate for other interpretations.

'The wetlands of Lefkada, Greece, inspired a body of work. Many sketches were made on several holidays and fascination for the location began to build in my mind. Ideas about composition and particular themes to exaggerate led to earlier work. Due to the Covid lockdown there was time to explore so many other aspects, and luckily my earlier sketchbooks provided reference for the growth rhythms, the variety of grasses, the water patterns, reflections, and in particular the huge colour changes witnessed each year as well as those affected by the time of day. I enjoyed highlighting certain aspects and negating others. To celebrate this place was a joy but underpinned by the sadness of knowing that ongoing climate changes could totally destroy this fascinating area forever.'

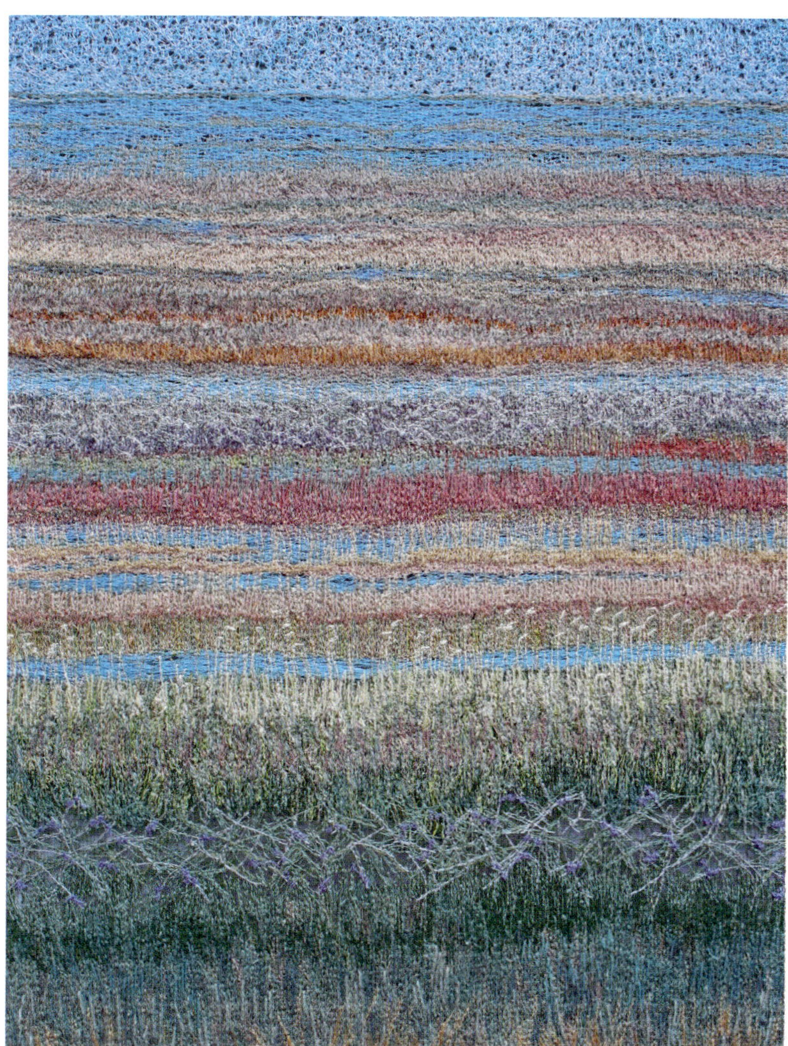

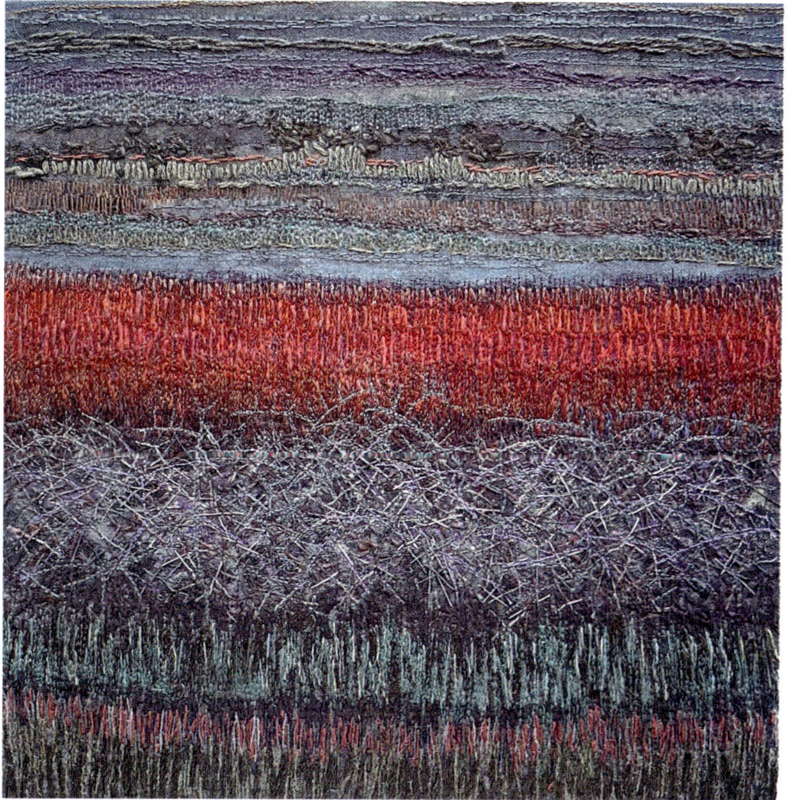

FEATURED ARTIST

Jean Littlejohn

'The works featured here, *Moon Pieces*, form part of a larger body of work developed over several years. They emanate from an interest in the natural cycles that influence and enrich our lives.

'The moon has held a compelling fascination since the evolution of human development, and many cultures have sought to quantify and understand its power. I enjoyed the research associated with these pieces. Each full moon has a name that varies according to culture and climate. They reflect the human experiences and animal behaviours associated with them. For example, on the night of the Worm Moon, worms are seen to rise to the surface of the ground. In January in Scandinavia they celebrate the Wolf Moon, whereas elsewhere this is often called the Snow Moon.

'Equally important to my work is the recording of visual elements, cloud studies and descriptions of night skies with careful notes on colours, mood and movement.

'I often use quotations as a source of inspiration. A phrase remembered from childhood, "the moon is a ghostly galleon tossed upon cloudy seas", from a poem by Alfred Noyes, was particularly apt.

'The final pieces are a synthesis of all these elements and are not literal. They are underpinned by observation but gather their own momentum as the work progresses. Elements may be exaggerated or played down according to the mood of the piece.

'I start with rough drawings, but once the basic composition has been established the simple straight stitch marks almost gather their own momentum and the work progresses organically.

'The techniques I have used are appropriate to this series and vary according to the subject matter. The Moons have been created from dark-coloured silk viscose velvet, mostly using recycled garments. Discharge paste was used for colour removal, applied with a sponge to produce a varied print. The background cotton has in most cases also been discharged to evoke a cloudy sky.

Below left: Jean Littlejohn, *Blood Moon*. Mixed media textile

Below right: Jean Littlejohn, *Tossed Upon Cloudy Seas*. Mixed media textile

Opposite: Jean Littlejohn, *Thunder Moon*. Mixed media textile

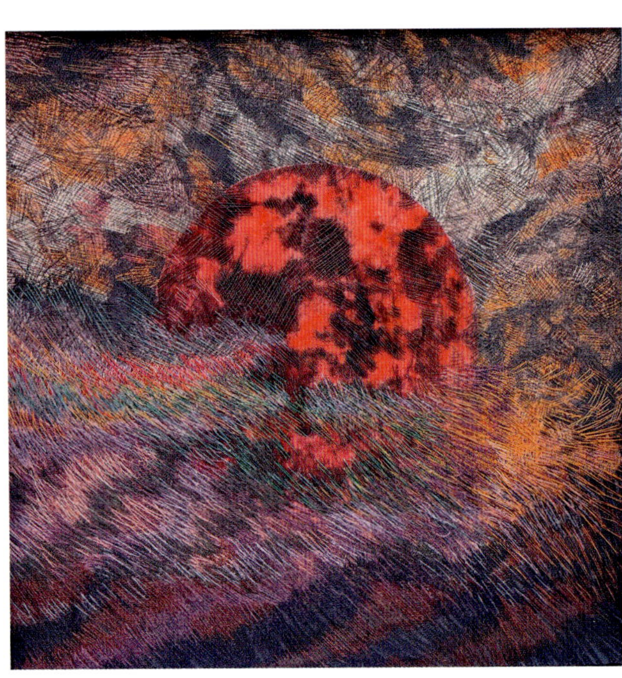

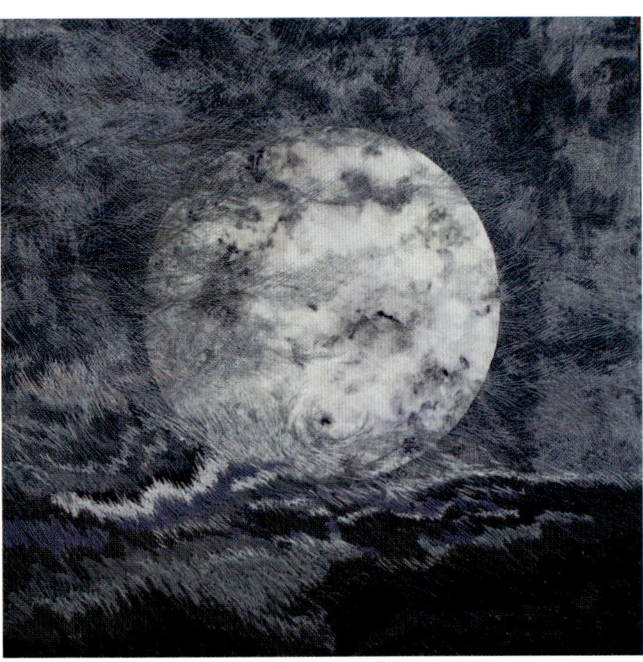

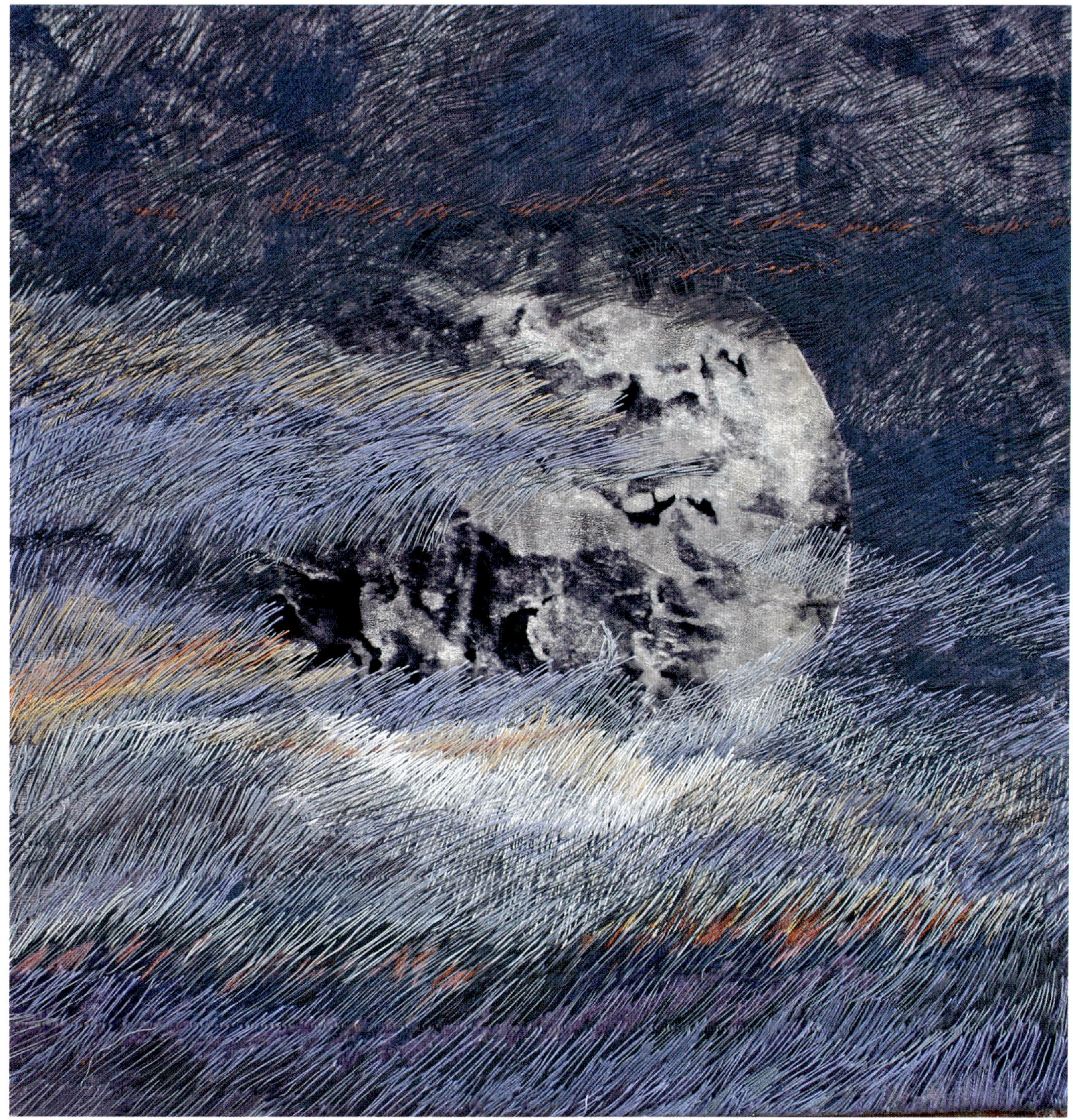

'Any surface elements such as moons, foliage and trees have been bonded on and the work completed with many layers of straight stitches to express depth and movement.

'I have loved the involvement with this series of work, and my advice to anyone starting is to work on inspiration that really engages you.

'Observational studies are so much better than photographs as they can remind you not only of how something looked but how it felt at the time. Of course photographs can back it up. It might also help to look up literary quotations and cuttings from newspapers and media sources where appropriate.

'Finding images from established artists in other disciplines might also be useful – not to copy, but to inspire. The most important thing is to keep looking and thinking and to record your thoughts.'

NATURE AND SAMPLING IN STITCH

The natural world provides a limitless supply of starting points for abstraction. Plants are a good place to start, as they have everything the textile artist needs: form, pattern, colour and texture. When I started machine embroidery, I made these simple plant embroideries, abstracting the shape and colour from drawings and photos. Often, working in a single colour can help to focus your concentration on the shape and patterns you choose to work from.

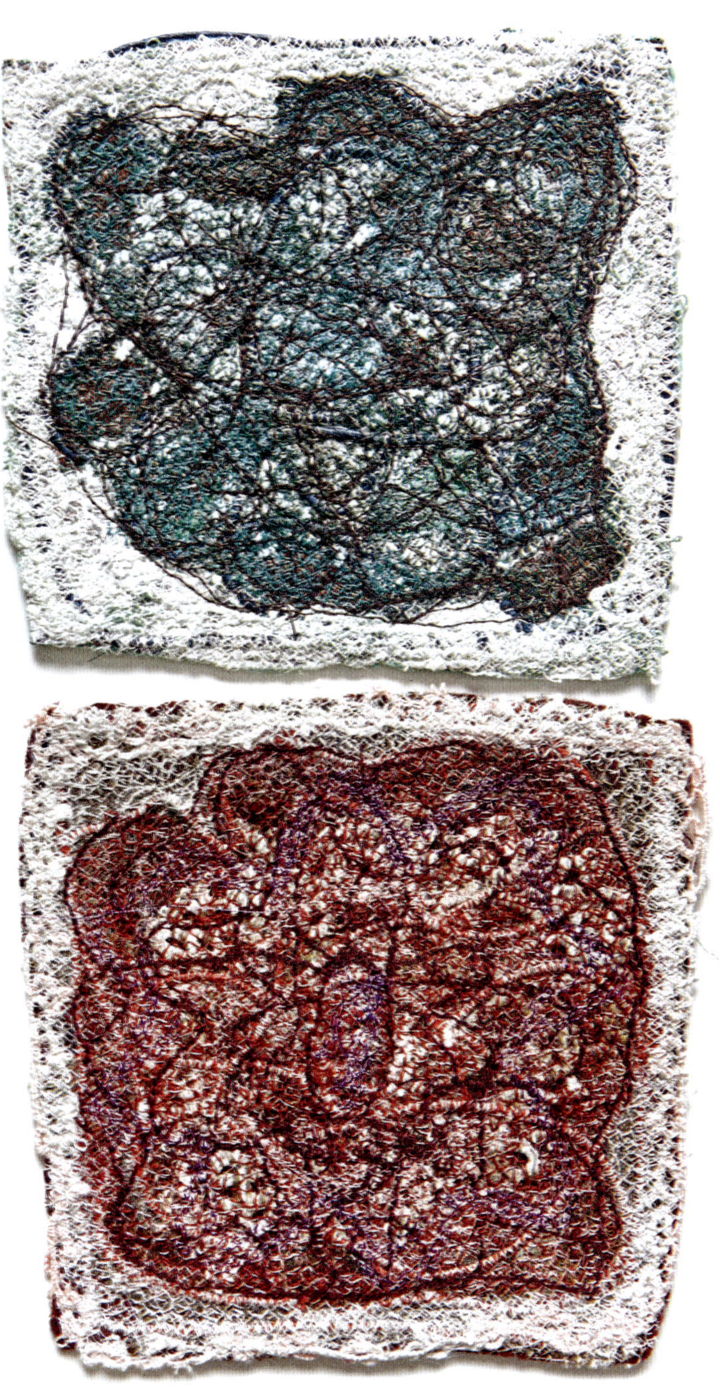

Left: Anne Kelly, *Small Square Plants*. Mixed media embroidery

Opposite top: Anne Kelly, *Small Square Plant*. Mixed media embroidery

Opposite bottom: Anne Kelly, *Small Square Plant* (detail). Mixed media embroidery

WORKSHOP

Line

MATERIALS REQUIRED

- Small strong fabric (cotton, linen, canvas) squares or rectangles for a base
- Tracing paper, greaseproof paper, pencil and/or Frixion pen
- Masking tape
- Hand-stitching kit
- Machine-stitching threads (if desired)

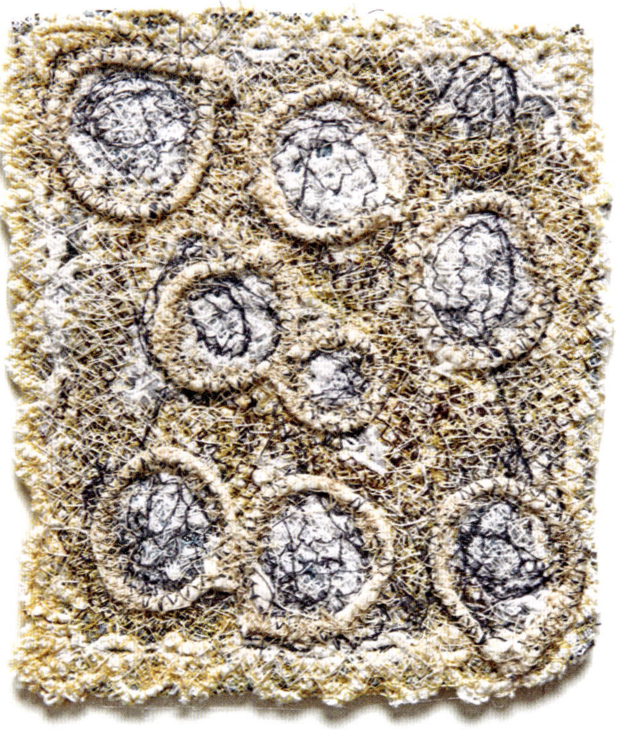

METHOD

1. Start with a photo or drawing from a chosen plant.
2. Isolate a section of your image using a viewfinder or frame.
3. Sketch the selected area onto a strong piece of fabric, using a pencil or Frixion pen.
4. Choose a thread colour that will be striking and expressive, according to your design.
5. Stitch the lines by hand, machine or both.

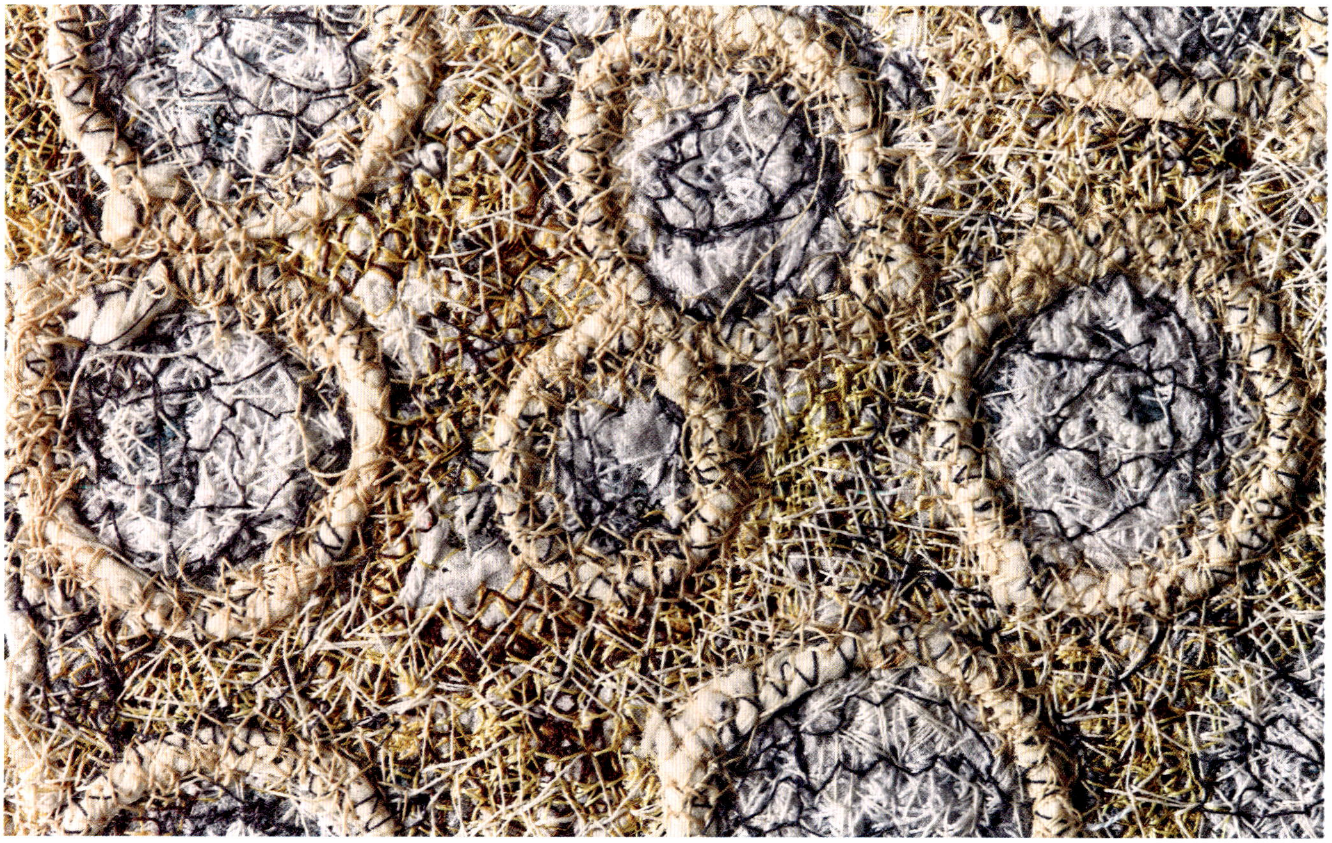

TEXTURE

Often the most striking element of an inspiring natural form can be its texture. Whether it is a barnacle-covered shell or a smooth stone, pitted by the sea, these surfaces can be manipulated to create a range of samples.

I like to assemble materials and media that remind me of the source imagery. Here are some materials that lend themselves to expressing texture in textile collage:

- Strong fabrics, tight or loose weave, that can be frayed and distressed
- String and hessian ribbon
- Scrim, binca, loose-weave needlework surfaces
- Non-stretch synthetic textiles such as organza and rayon
- Wool and heavy embroidery thread for stitching

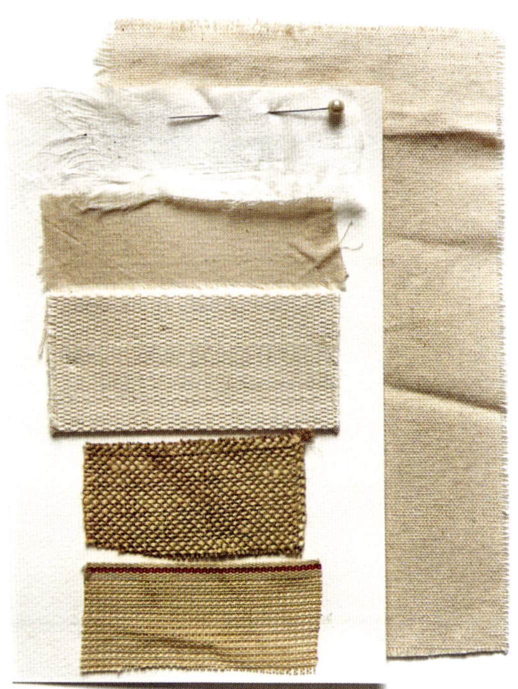

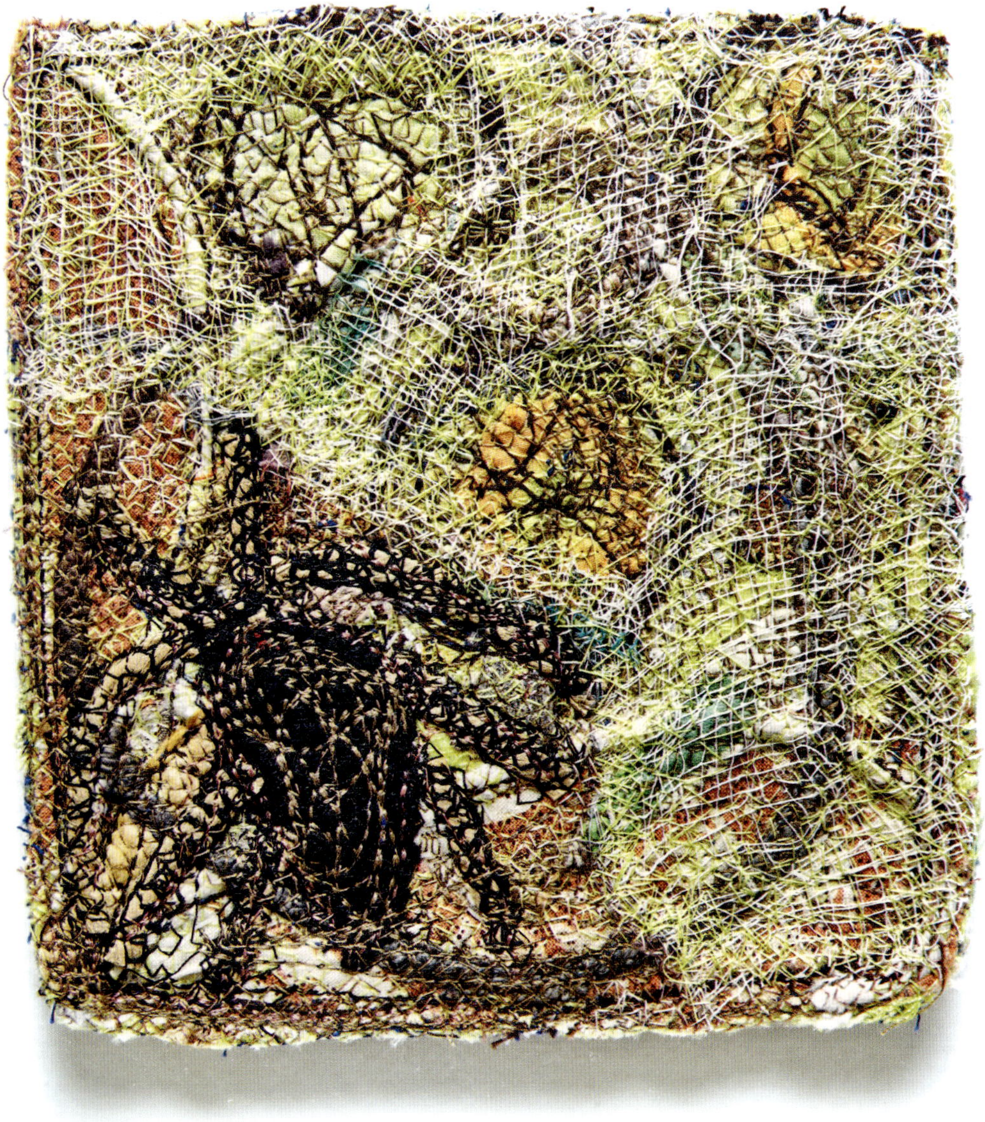

Above: Anne Kelly, materials for sampling with stitch

Left: Anne Kelly, *Spider*. Mixed media textile

WORKSHOP

Texture

MATERIALS REQUIRED

- Small square or rectangular scraps for sampling of strong cotton/linen cloth
- Plants, leaves or sections of them, rocks or shells flat enough to print with
- Fabric paint (or acrylic paint) or dye to colour backgrounds and for printing
- Foam or blanket cloth for cushioning the print
- Brushes, sponge (small)
- Masking tape

Below left: Leaf on calico for printing

Below right: Anne Kelly, *Leaf Print*. Acrylic on dyed canvas

METHOD

1 Choose some colours for backgrounds and dilute your fabric paint or use a ready-mixed dye to colour the cloth pieces.

2 When they are dry, select a range of printing materials, ensuring that they are flat enough to make an impression.

3 Tape your fabric to the foam or blanket so that it won't move.

4 Gently daub the object to be printed with a thin layer of fabric paint.

5 Press into the background and continue the process until you have filled the fabric.

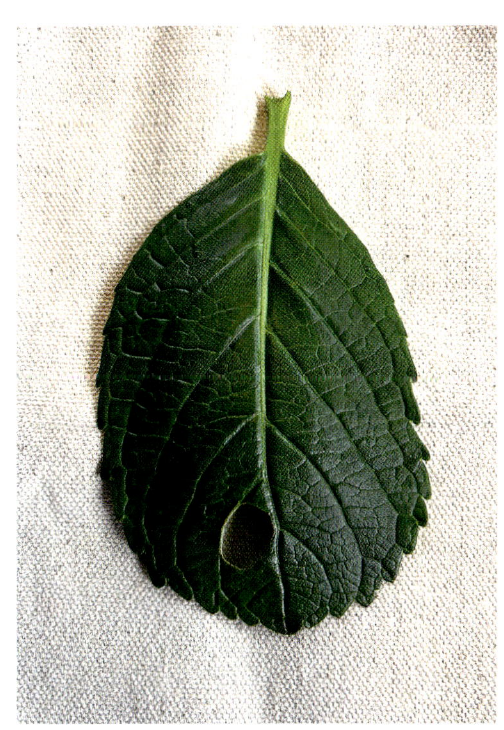

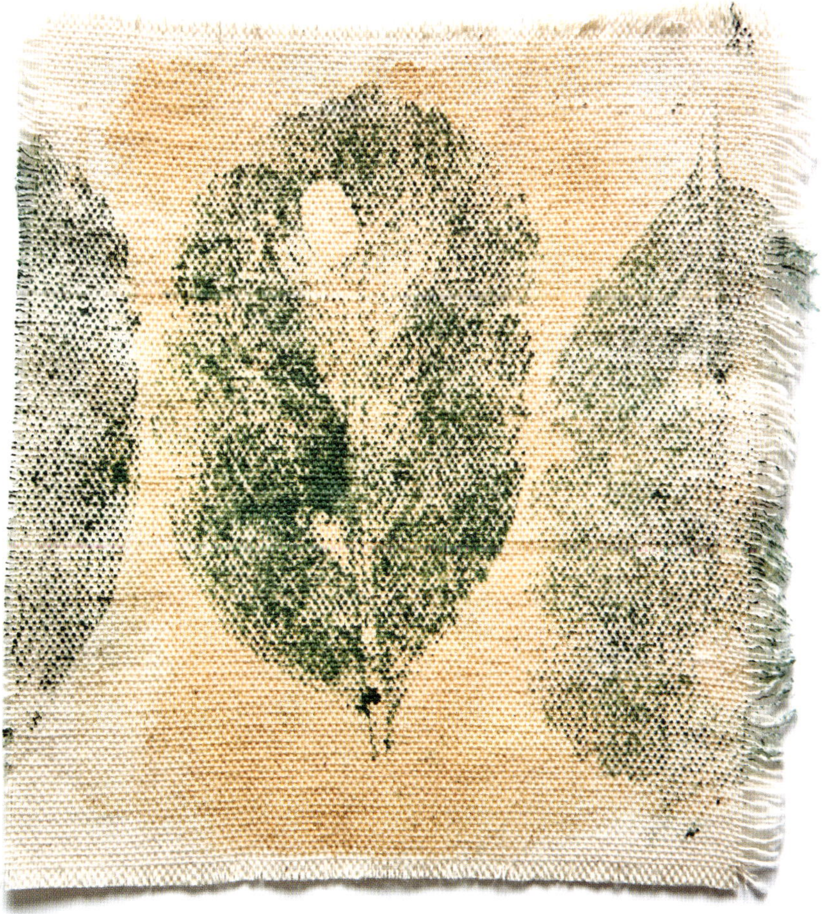

FEATURED ARTIST

Tara Axford

Tara Axford is an Australia-based artist and tutor. She shares her numerous and ever-changing experiments with a large online audience. I met her while teaching in Australia in 2019 and it is great to see her continuing her explorations. She tells us:

'When I look at the work of artists I admire, often there seems to be a strong sense of place – a big space conveyed, possibly a horizon line or implied distance. For me, most of what I see and experience on my walks is often visually messy and chaotic. The paths and trails are not always clear. The one thing that seems to be constant are the rocks: solid, majestic, sculptural, textural, the colours changing in the light. Another thing that delights me is the plant life – from the eye-catching banksias, grevilleas and gum blossoms, all evolving throughout the year, to the small species at ground level.

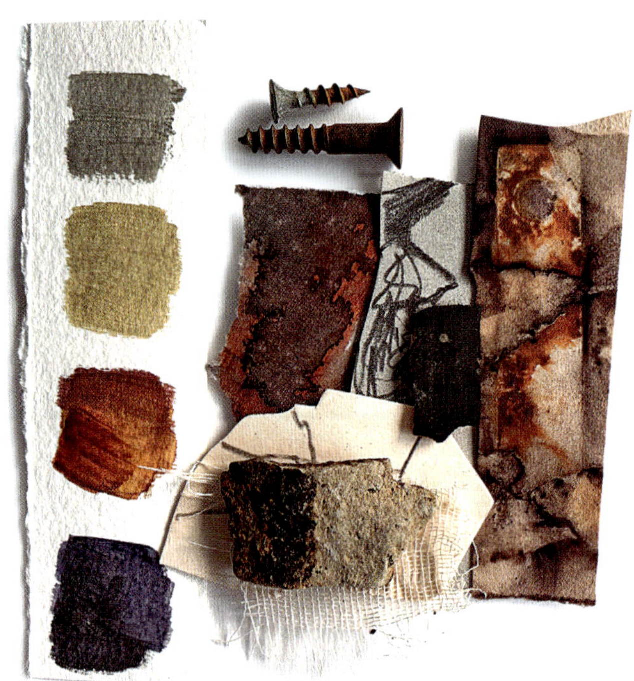

'These are the things that are familiar to me and those are the elements that I seek to "re-present". I see them all the time, I collect them, observe them, rearrange them, deconstruct them. I don't attempt to recreate them; rather, I aim to have them in my subconscious and have them flow out in other ways, trusting that their influence will appear in my process. I like to explore fragments, break down the elements, and make colour swatches from collected pieces.

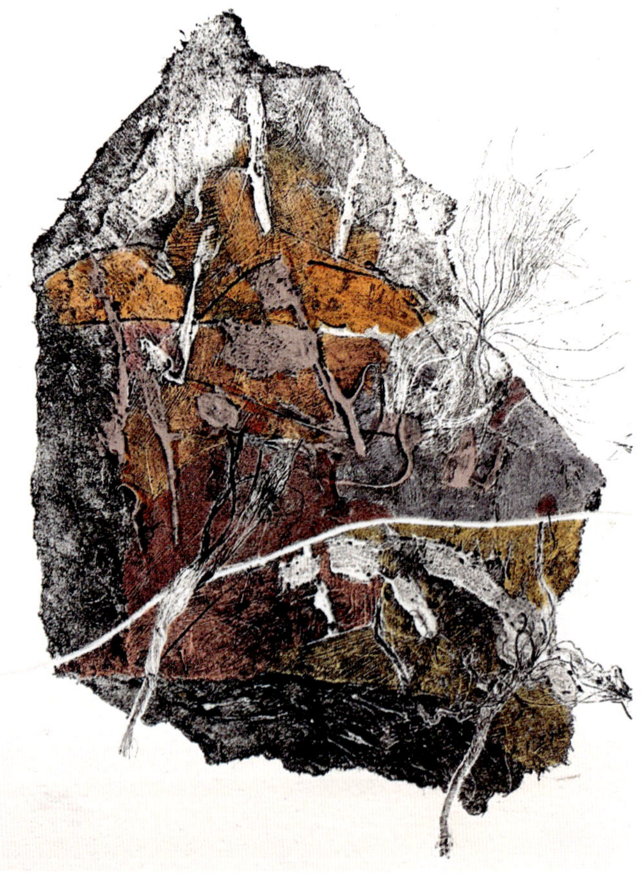

'The textures and marks created on my collagraph plates aim to capture the essence of what I have seen, but more importantly, what I have experienced. I also like to incorporate natural elements into the prints, the unknown. They permeate my consciousness, becoming a part of my artistic process – not to be replicated, but to be internalised and expressed in myriad ways. Fragmented explorations, deconstructed elements, and colour swatches crafted from collected specimens all find their place in my creative journey.

'Through the textured imprints of my collagraph plates, I endeavour to encapsulate not just what I've seen, but what I've felt. Incorporating natural elements into my prints adds an element of spontaneity. I find solace and inspiration in this interplay between the known and the unknown, a testament to the boundless wonders of nature and its endless capacity to ignite the artist's soul.'

Opposite top:
Tara Axford, *Quarry*. Mixed media

Opposite bottom:
Tara Axford, *Rock*. Hand-coloured collagraph print

Right: Tara Axford, *Solid Rock*. Digitally coloured collagraph print

Below: Tara Axford, *Shelfie 7*. Mixed media, found objects and prints

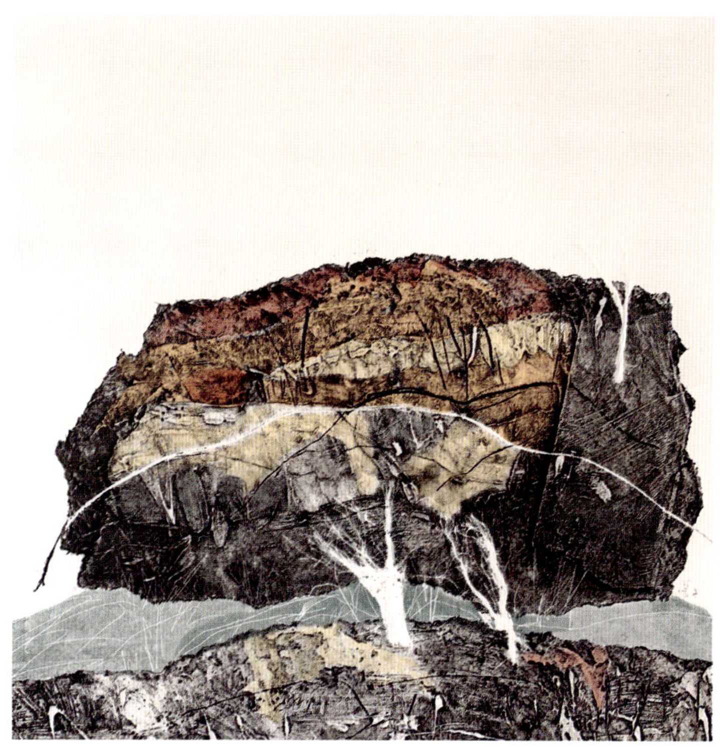

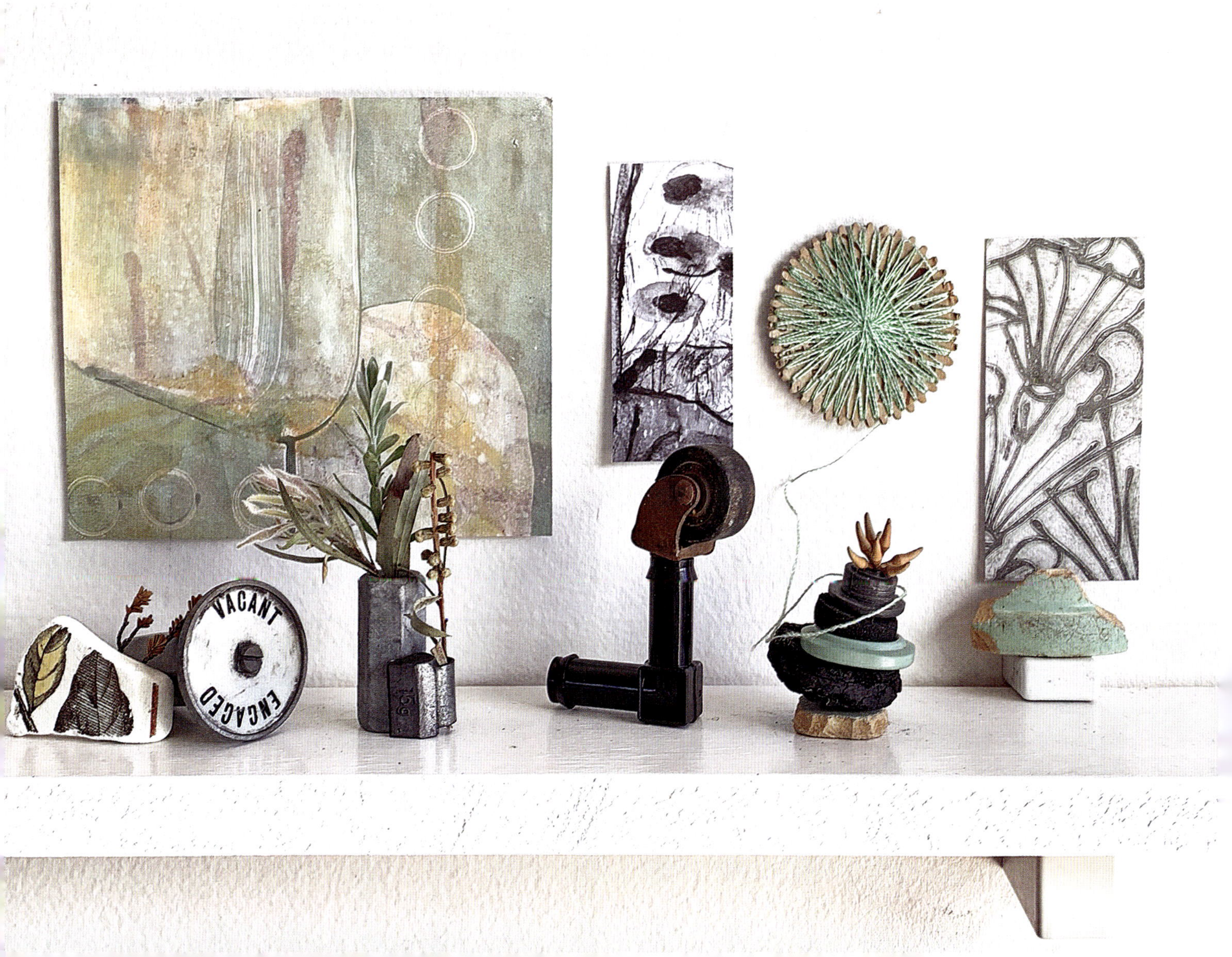

PATTERN AND NATURE

As we have seen in this chapter, nature can be a rewarding and satisfying 'way in' to abstraction. Using pattern and simplifying shapes makes the transition easier and more organic.

In my sample below, I have used a repetitive shape to create an overall pattern. The stitching and linear elements also contribute to this.

In my portrait work in progress, I have used all over patterns to indicate and delineate the different areas of the composition. Instead of 'making a picture', I am combining shapes, colours and textures.

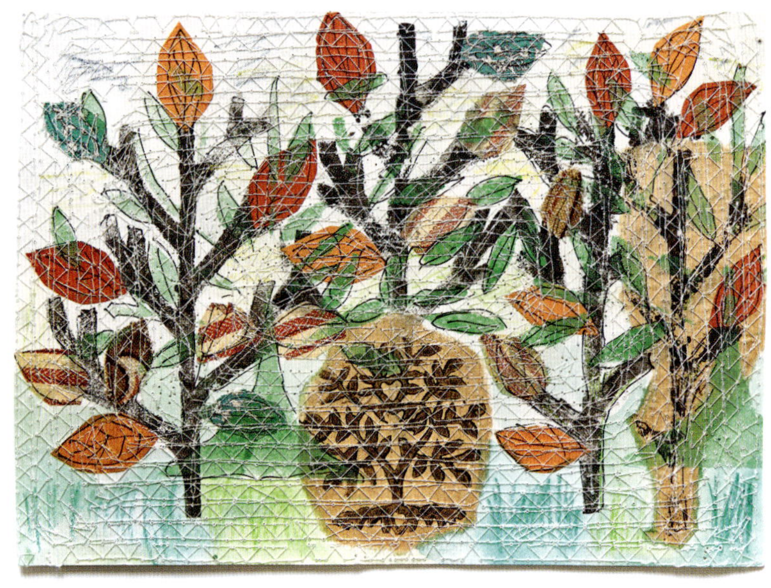

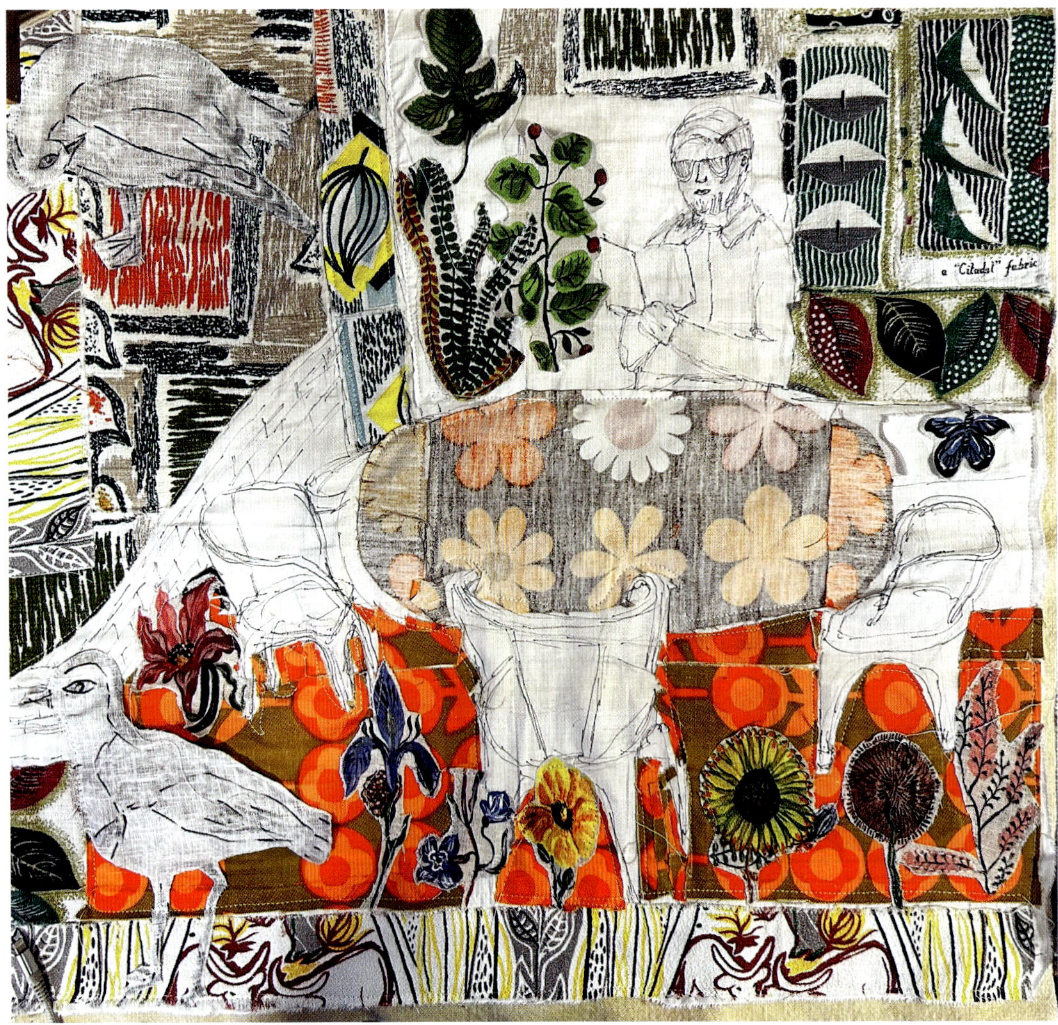

Above: Anne Kelly, *Tree Collage*. Mixed media textile

Left: Anne Kelly, *Garden Collage*. Mixed media collage

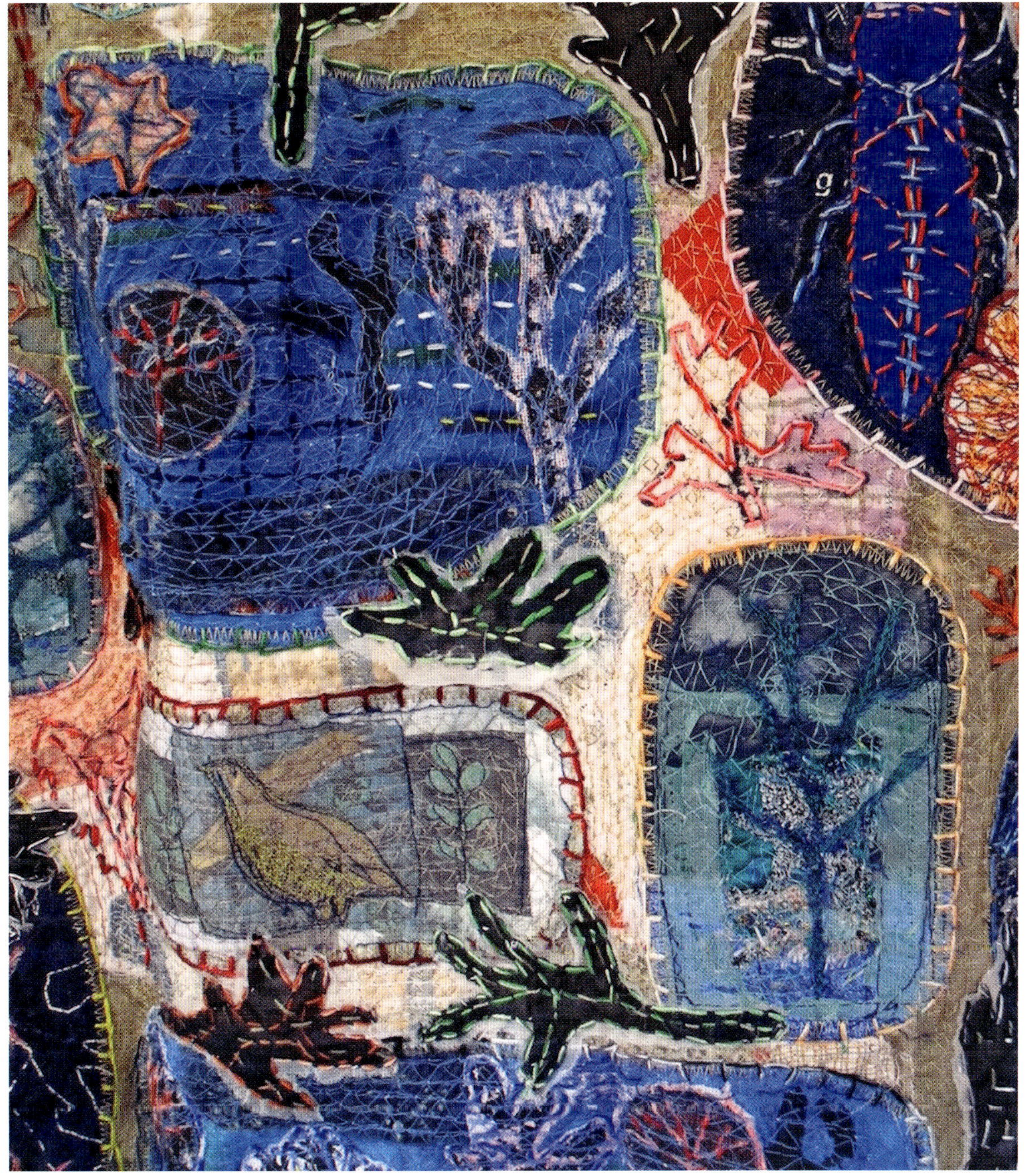

THE SEASIDE – MARKING SURFACES

My piece *Undersea* was a key component of my Well Travelled exhibition at the Ruthin Craft Centre and toured to the Harley Foundation in Welbeck. It was also selected for the Sussex Contemporary exhibition in Brighton in 2023.

I used rockpools and seashore images as the inspiration for this large circular piece. Each 'pool' started with a printed piece of fabric and a selection of vintage scraps, layered into the composition. I used hand and machine stitching to join the pieces together and emphasise the main outlines. I exaggerated the colours to create a more striking impression. The whole piece was mounted on a vintage tablecloth.

Above: Anne Kelly, *Undersea* (detail). Mixed media textile

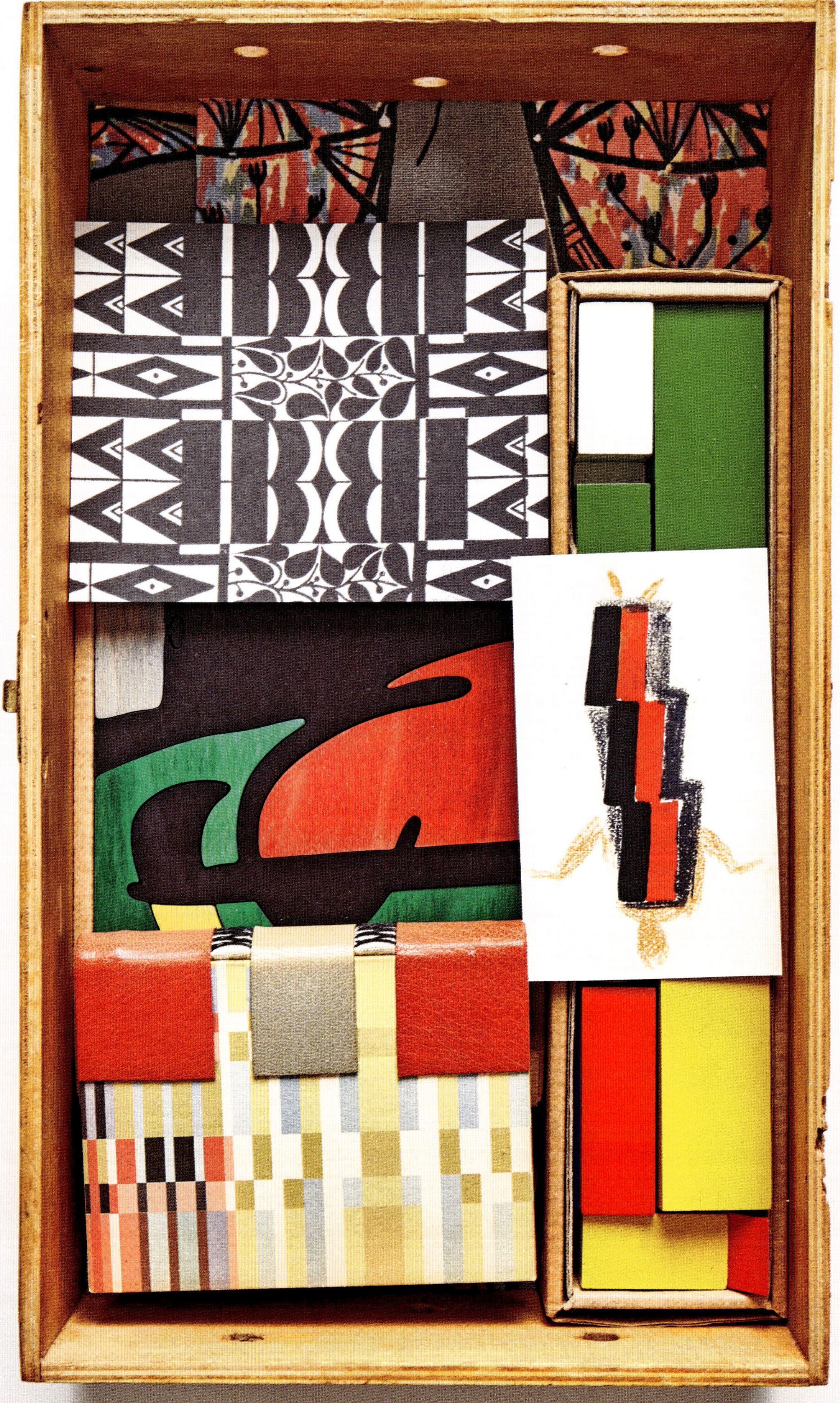

Chapter 2

SHAPES AND COLOUR – MID-CENTURY INFLUENCES

'The longer you look at an object, the more abstract it becomes, and, ironically, the more real.'

Lucian Freud (1922–2011)

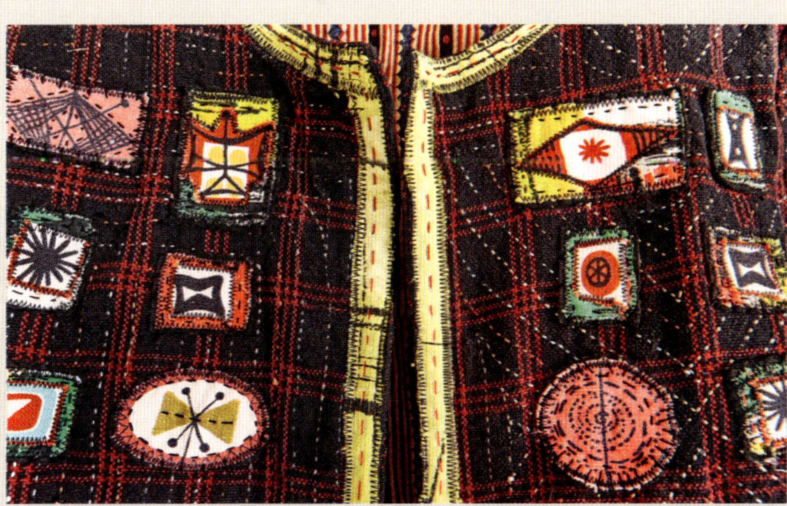

Opposite: Anne Kelly, *Inspiration Box, Mid-century*. Mixed media in wooden box

Right: Anne Kelly, *Mid-century Meadow Jacket* (detail). Appliqué and embroidery on textile

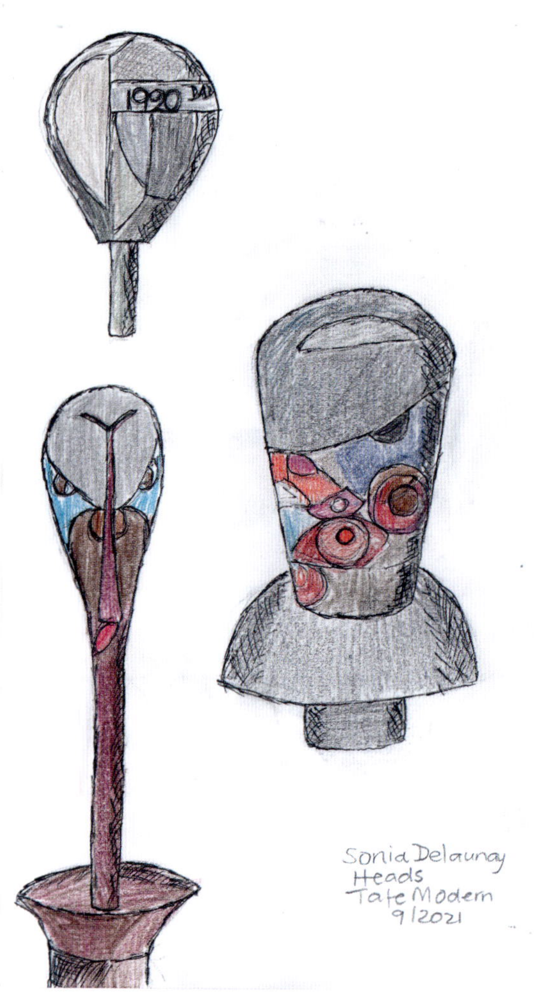

There is something at once comforting and challenging about mid-century art and design. It represents a portent for the future – a vision of symbolic and simplified shapes – and yet we can see recognisable elements in it. The scale is human, and these artifacts often fit well into our modern and post-modern lives.

Shape and colour take centre stage in this chapter. We explore how we use these shapes, in two- and three-dimensional forms, by painting and dyeing. Houses and house imagery can be a link between the narrative and the abstract, like a bridge. Simple mid-century block printing techniques will be covered too.

Being inspired by the shapes and simplified forms of this area of art can lead us into other areas of study, so patchwork and appliqué are discussed here. Using vintage and mid-century textiles can also lift and inspire your projects. I will show you how to select them and use them in new ways.

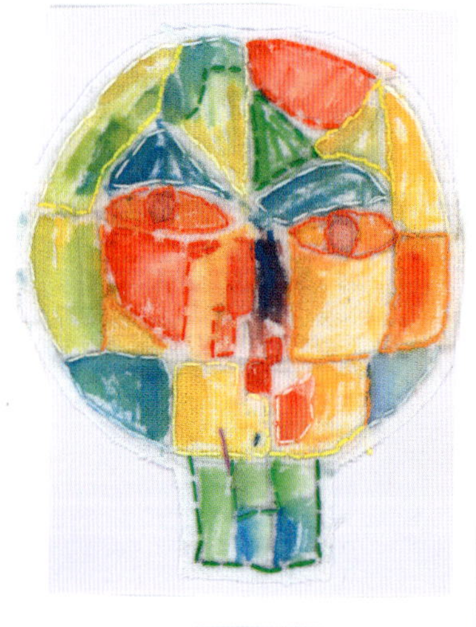

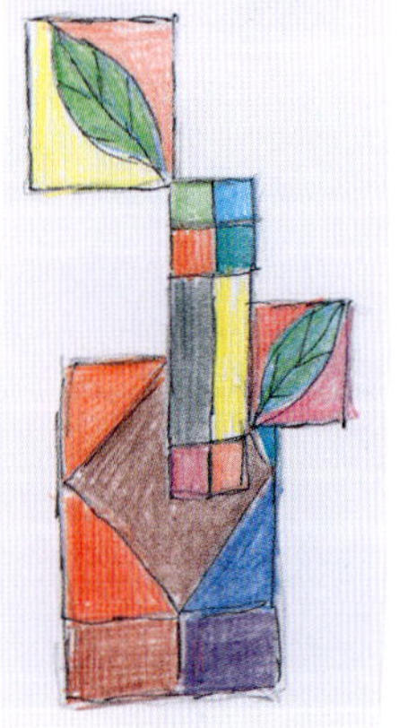

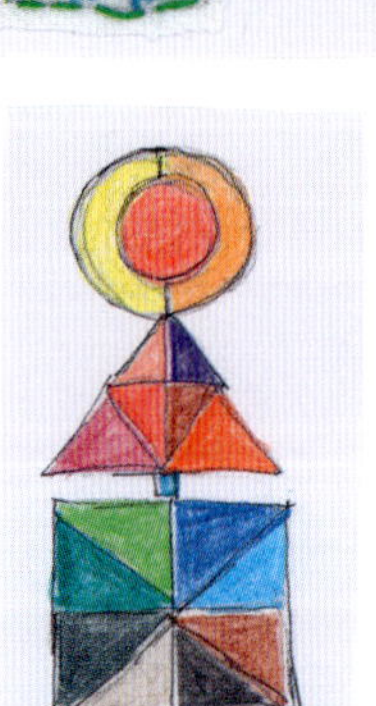

Geometric Samples

Left: Anne Kelly, *Sketchbook Pages*. Mixed media

Opposite top: Anne Kelly, *Painted Patches*. Fabric paint on canvas

Opposite bottom: Anne Kelly, *Patched Heads*. Mixed media textile sculpture

WORKSHOP

Making a patched head sculpture

Inspired by Sophie Taeuber-Arp's exhibition at Tate Modern in 2021, I decided to make a patched head sculpture. I looked at her drawings and paintings for the design of it and adapted them to make one in her style.

MATERIALS REQUIRED

- A mannequin head for hats, any material
- Canvas or heavy cotton squares
- Fabric paint (or acrylic paint)
- Paper for pattern making (greaseproof paper is good)
- Machine- or hand-stitching thread for sewing
- Pins with heads
- Glue gun and glue sticks
- Fabric to cover head if necessary

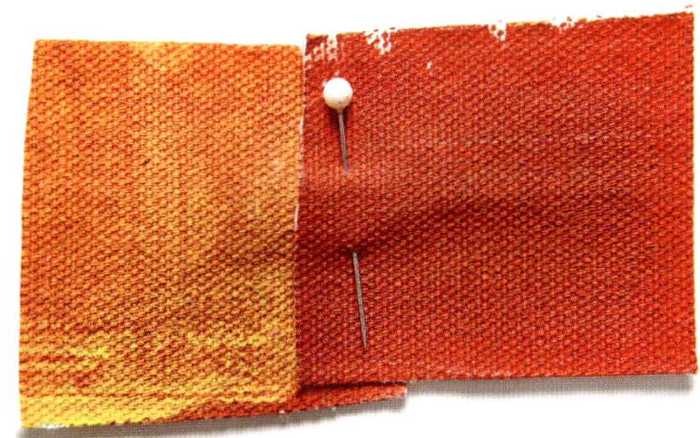

METHOD

1. I painted scraps of canvas in primary shades with fabric paint. When they were dry, I stitched them into strips on the machine and made enough to cover the head.
2. The head is a hat-making mannequin head on a stand. It was covered in cotton jersey.
3. I made a pattern from paper to cover the head. I started on the front and covered it over the entire shape, following on with different strips until the head was covered.
4. I attached the pieces with pins and then glue-gunned them into place with small dots of glue.

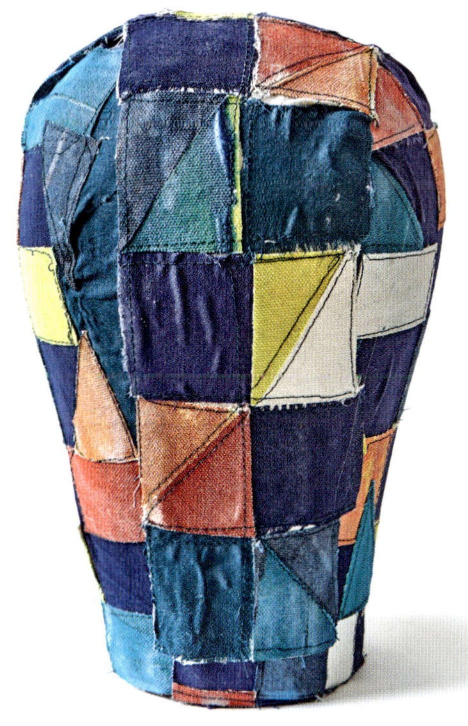

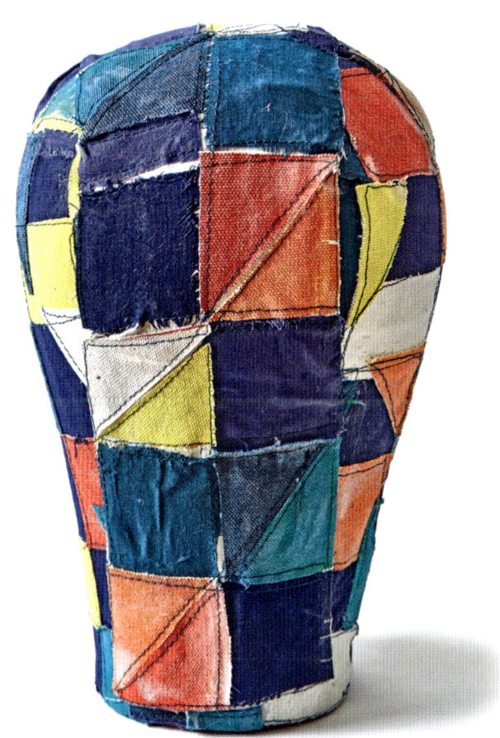

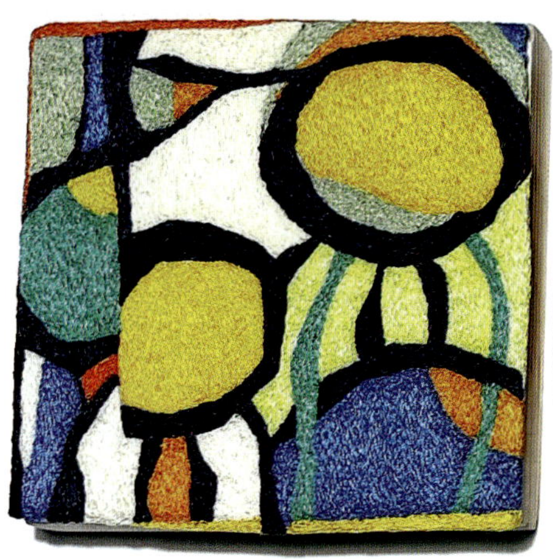
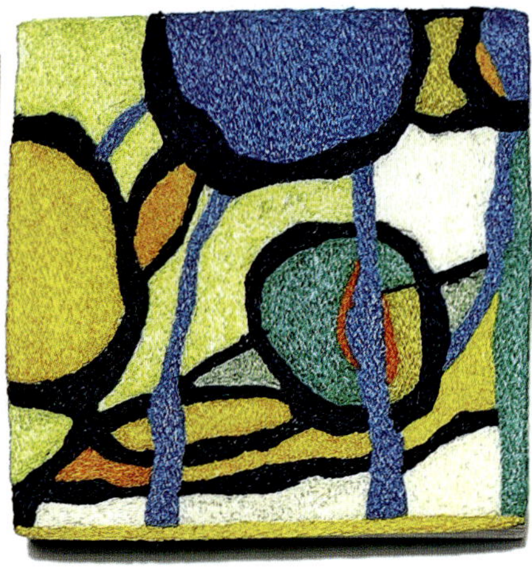
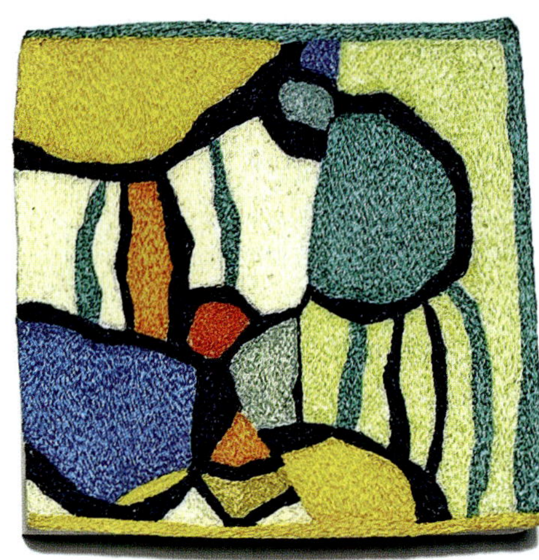

FEATURED ARTIST

Helen Banzhaf

Helen Banzhaf is a well-established embroiderer who makes striking geometric-shaped pieces. She describes her practice and inspirations:

Above: Helen Banzhaf, *Three in a Row*. Unframed machine embroidery on wooden block

Below: Helen Banzhaf, *Work in progress*. Machine-embroidered neckpiece

Opposite top: Helen Banzhaf, *Neckpiece*. Machine embroidery

'My whole creative life has revolved around designing, making and the joy of working with stitches. Cloth and threads are the fundamentals in nearly all of my creative activities.

'Just over thirty years ago, while working as a fashion lecturer, I began exploring using these mediums to decoratively stitch pictures onto calico "canvases" using the technique of free machine embroidery. I work on a domestic sewing machine; one I also use to make my clothes.

'I have a collection of 1930s Art Deco ceramics and my seminal embroidered piece was based on a jug of which I was particularly fond. Here began my stitched decorative textile journey!

'Thereafter, my penchant for vessels, their shapes, their forms, their colours and patterns, became my predominant subject matter. I found them inspirational and they have recurred in my work over and over again. The ceramic pieces can become distorted and they float in space with the decorative pattern mingling in and around them. There is still an element of representation but with hints of an abstract quality.

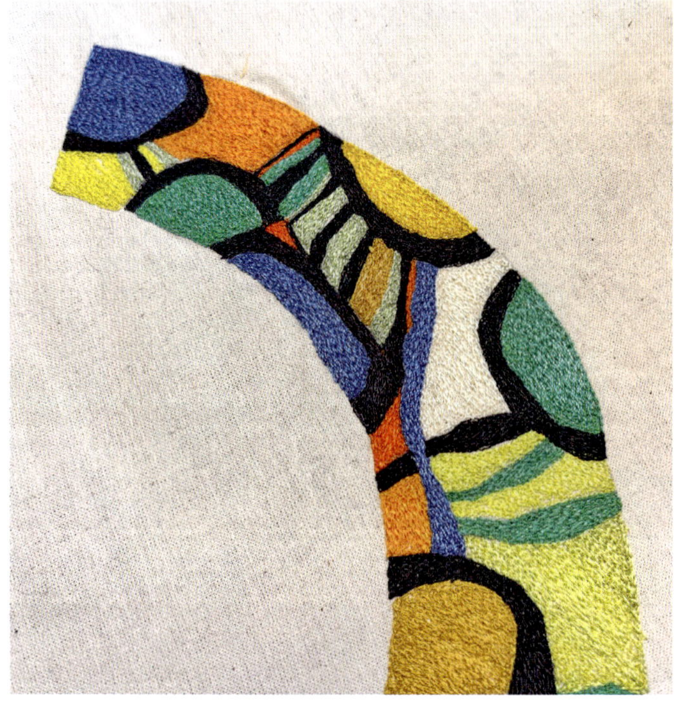

'The abstract element comes by sometimes unexpectedly. The starting point and design development isn't a terribly natural journey for me, but once I think I've cracked it, I'm on my way. I admire abstract art hugely. In the early days I was influenced by, for example, the paintings of Sonia Delaunay; her vibrant use of colour and pattern, and the clarity of her interlinking shapes was, and is, inspirational. Art and design in its many formats during that era still excites me. I am a fan of the work of Jean Hélion, Roger Hilton, Nicolas de Staël, Roger Cecil and the St Ives School of Artists, for example.

'My recent series *A Construction of Old Friends* was based on some of my favourite jugs, pots and bowls, which I arranged in a higgledy-piggledy still-life mound so that the individual ceramics were indistinguishable. The resulting embroideries were semi-abstract tumbles of shapes, patterns and colours. By playing around with the already slightly abstracted *A Construction of Old Friends* designs, I was inspired to explore and develop wholly new and out of the blue patterns and shapes with their very different

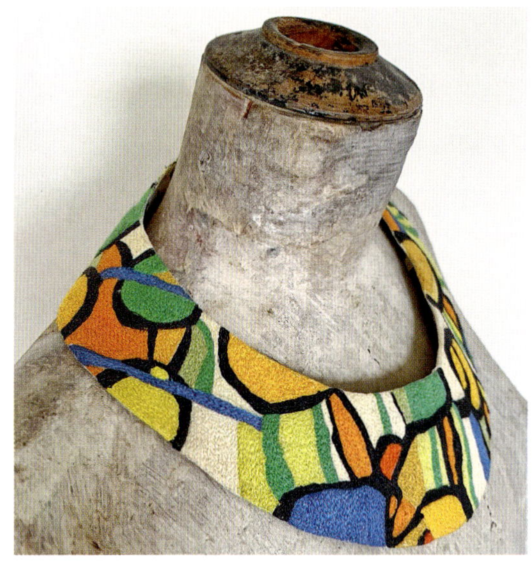

colour combinations. This is always a fun exercise and the triptych *Three in a Row* and the beginnings of the embroidered neckpiece seen here are the results of one of my enjoyable textile design and making journeys.

'Sometimes I will hone in on a small section of a bigger finished piece. I will take that snapshot image and work with it to develop a design that ultimately will have no recognisable relationship to the original embroidered piece. It becomes an abstracted pattern of densely worked stitches in its own right.

'My recent embroideries are presented as triptychs on wooden blocks without glass or frames. I designed my first triptych after a protracted creative paralysis during the Covid period. The design grew without much thought and minimal preparation. It happened organically and I was able to evolve the imagery, and with renewed gusto, much to my relief! In among each of the three stitched patterns there are always outlined, squashed, misshapen circles or ovals. It's a motif that has cropped up in my work over all the years.

'I enjoyed this set of stitched abstract "landscapes" and its colour themes so much that I explored the design further and used its essence as the decorative design for the embroidered neckpiece.'

STILL LIFE COLLAGE

Simple shapes and familiar objects can often provide inspiration for textile collages. Here I used the shape of an old vase to create the main feature in the composition. The background is made from vintage tea towels and other remnants. The vase is free-motion stitched on top of the collage. The entire piece was mounted onto a canvas board.

Below left: Anne Kelly, *Textile Collage*. Mixed media textile

Below right: Anne Kelly, *Textile Collage*. Mixed media textile

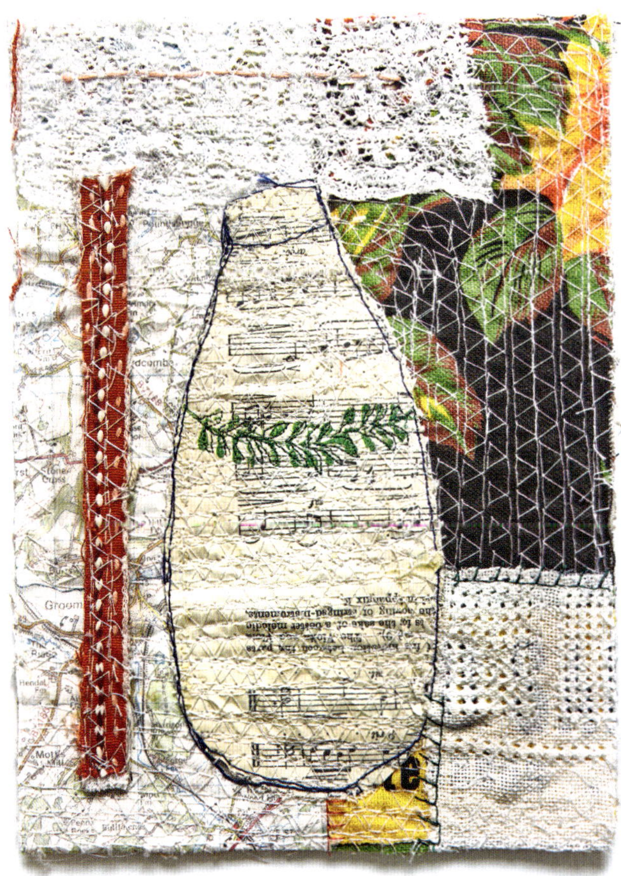

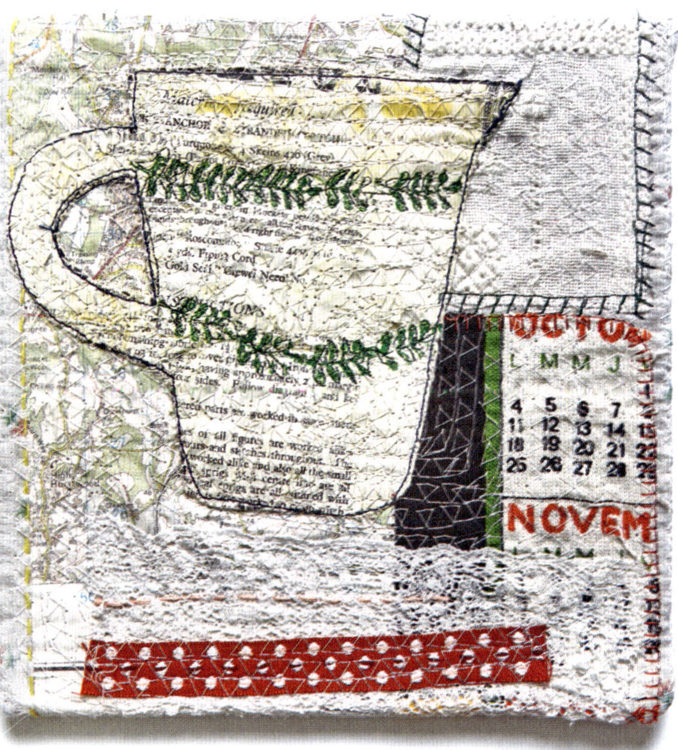

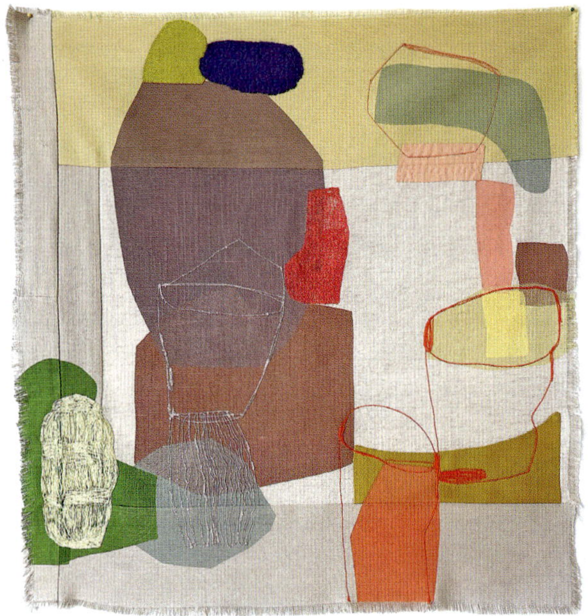

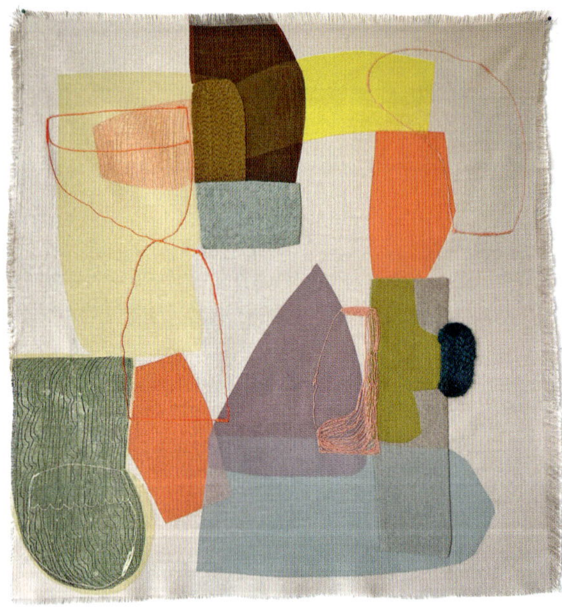

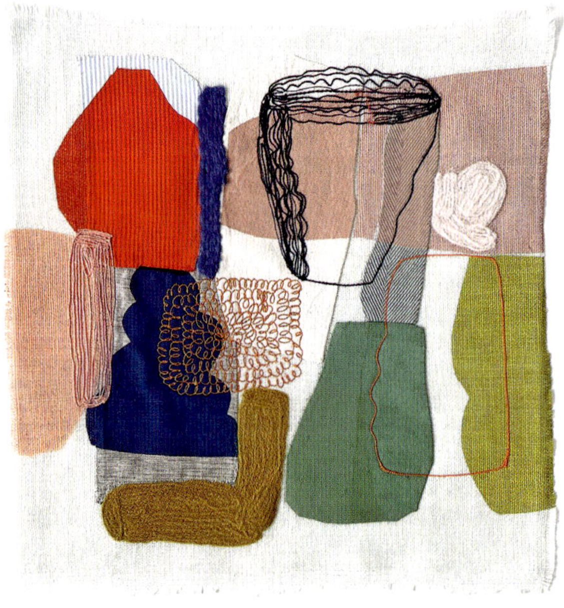

FEATURED ARTIST
Maxine Sutton

Maxine Sutton makes artworks using a range of materials and techniques. Her work strikes a balance between organic shapes and sensual textures. She talks about it here:

'Early experiences of making with my grandmother keep me connected to textile materials and needlework techniques, which are now layered with my later experiences of studying fine art painting. It was at art school in the 1980s that I discovered abstraction and immersed myself in the craft of painting large canvases inspired by Abstract Expressionist painters such as Arshile Gorky.

'My current practice combines the domestic making of my childhood experiences, a love of textile crafts, design and folk art, with my endless fascination for colour interactions and abstraction, all absorbed through my formal education in contemporary painting and printmaking. Employing a visual language hovering between abstraction and representation, the work plays with articulations of space, object and figure. Fragmentation and restoration, women's stories and ecology are some themes that permeate my thinking and making processes.

'I am interested in expressions of restoration and repair and how to make sense of the fragmentation of contemporary life. My work is underpinned by an interest in the environment, questioning how we value materials and what we choose to throw away. I use found fabrics, yarns and threads that often have a domestic and vernacular history. The echoes of everyday life present in these materials are layered with abstract shape and line, accessing emotional responses to materials and forms and looking for ways to express the unseen aspects of our human experience.'

Top: Maxine Sutton, *Memory Tray*. Screen print, appliqué and embroidery on linen

Centre: Maxine Sutton, *Where Shall We Go*. Screen print, appliqué and embroidery on linen

Bottom: Maxine Sutton, *Heeled*. Screen print, appliqué and embroidery on linen

MID-CENTURY TEXTILES

I have always had a soft spot for mid-century textiles – I find them nostalgic, as they remind me of the fabrics of my childhood. Through a fortunate encounter with Wendy Shaw, a collector of vintage textiles, I have been able to source some wonderful examples to use in my work. They are fragile and delicate due to their age and usage, but can be incorporated with gentle stitching and protection with a backing fabric. I like to use them for colour and texture, as they have so many rich patterns and colours printed on them. The texture of the fabric itself also adds a lovely vintage feel to the work.

Below: Selection of vintage mid-century textiles, sourced by Wendy Shaw

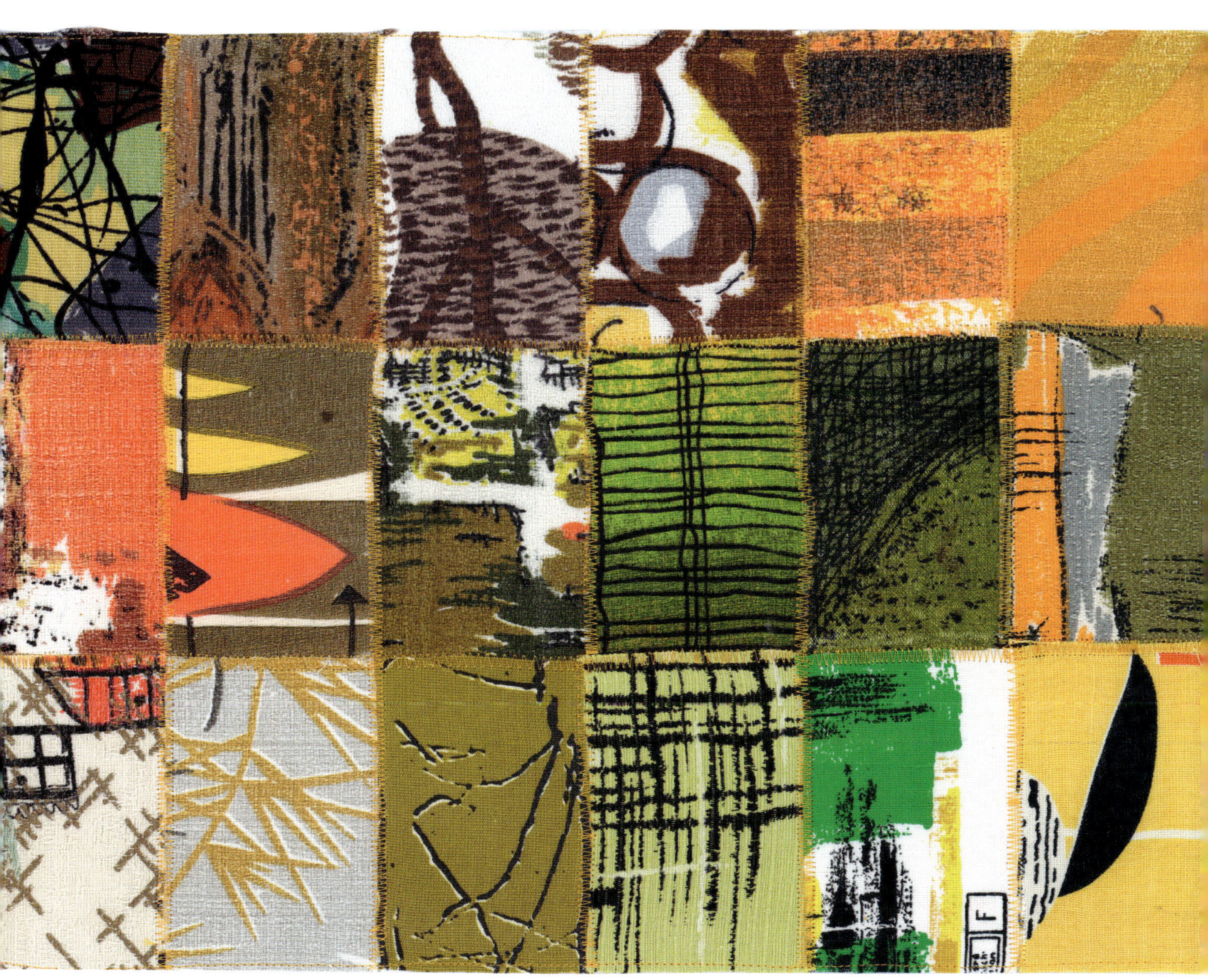

MID-CENTURY COMPOSITIONS

In these pieces, I started with a photograph and scaled it up onto a blank piece of vintage cotton and linen blend. I then used the mid-century fabrics to make the collage, attaching them onto the background with stab stitching. The pieces of fabric were then overstitched by hand and machine, to add texture and to bring the piece together. The areas of contrasting colour create new patterns and frame the composition. Finally, hand stitching adds definition and outlines the piece.

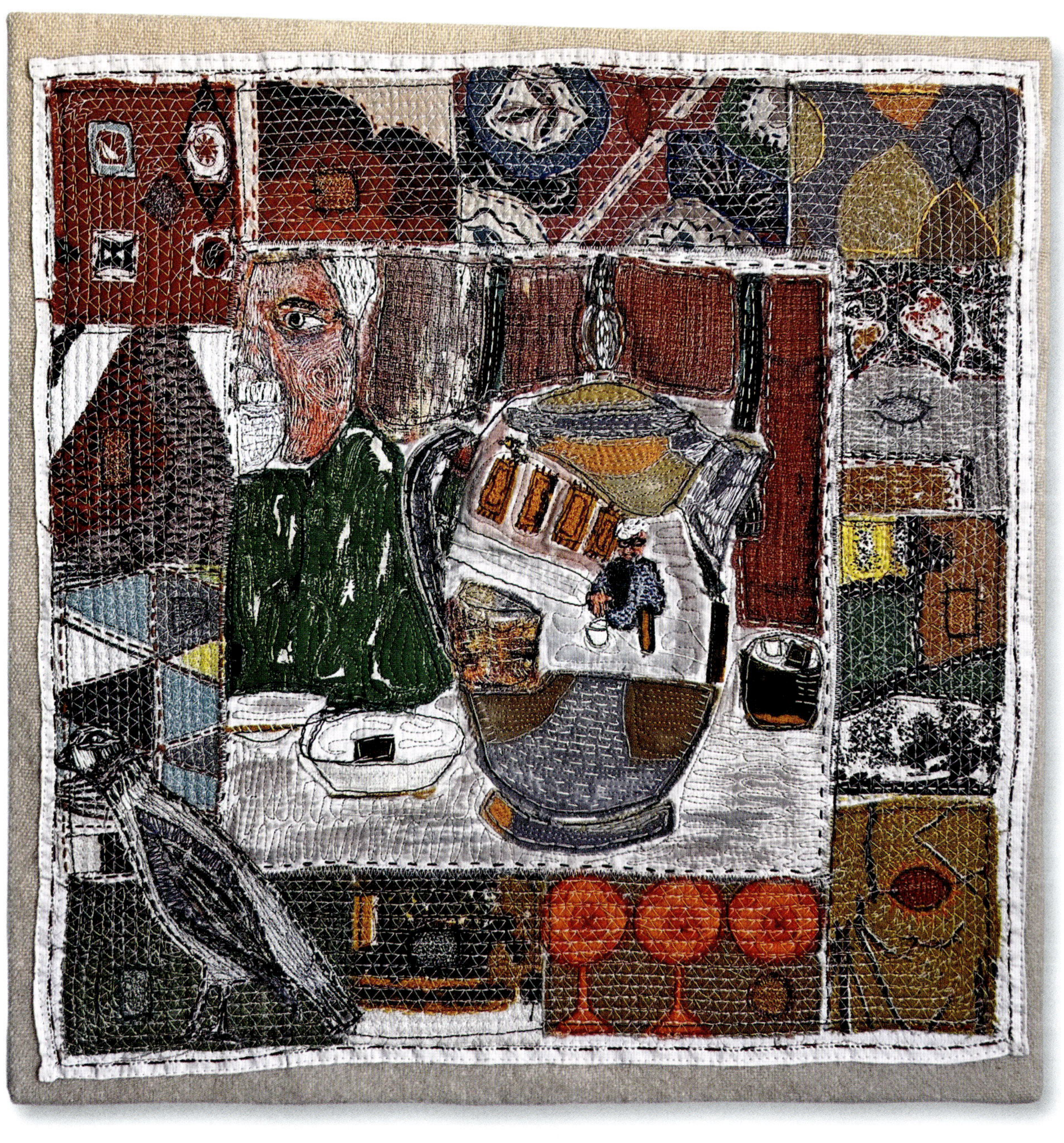

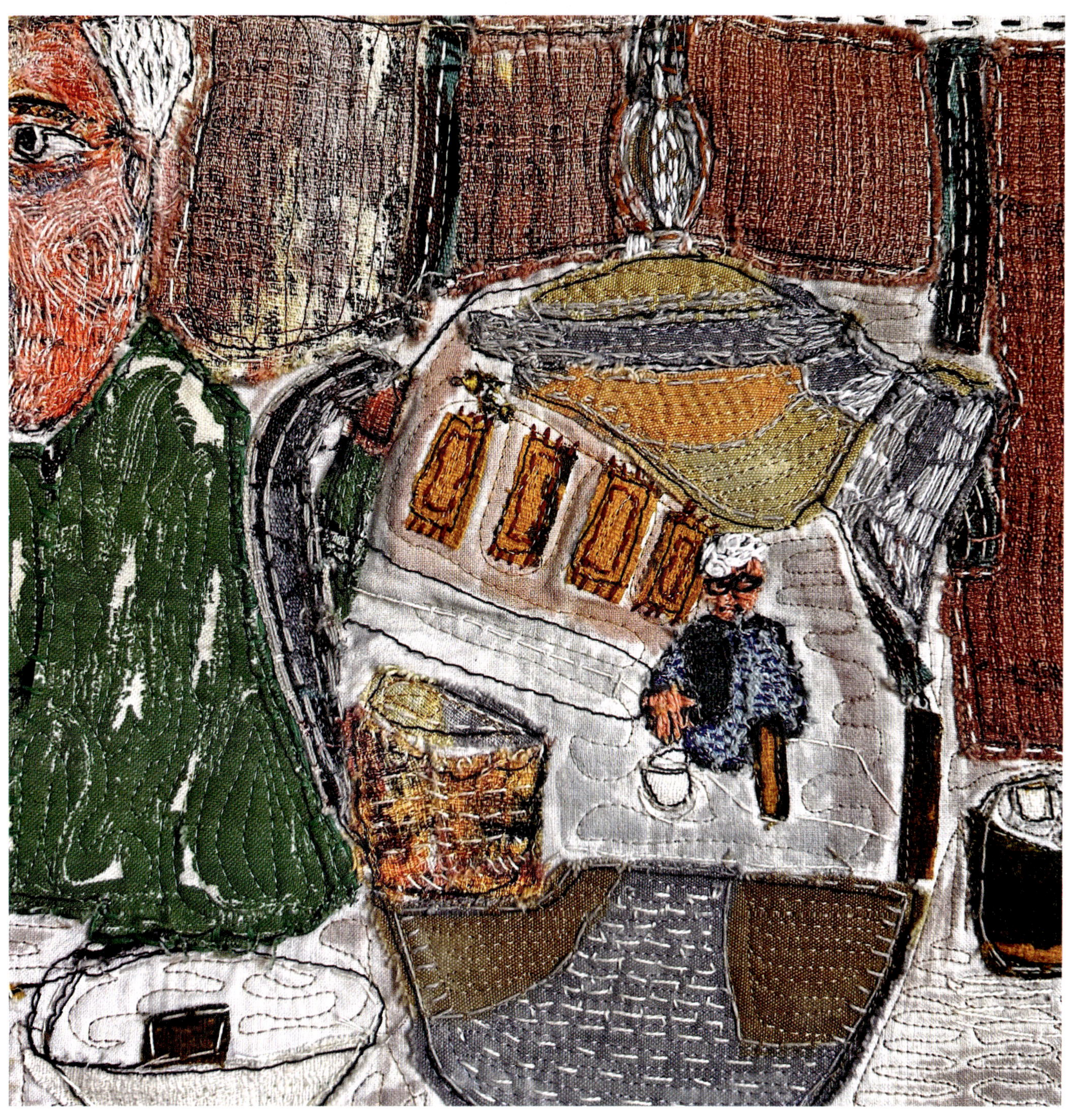

Opposite: Anne Kelly, *Reflections*. Mixed media textile

Above: Anne Kelly, *Reflections* (detail). Mixed media textile

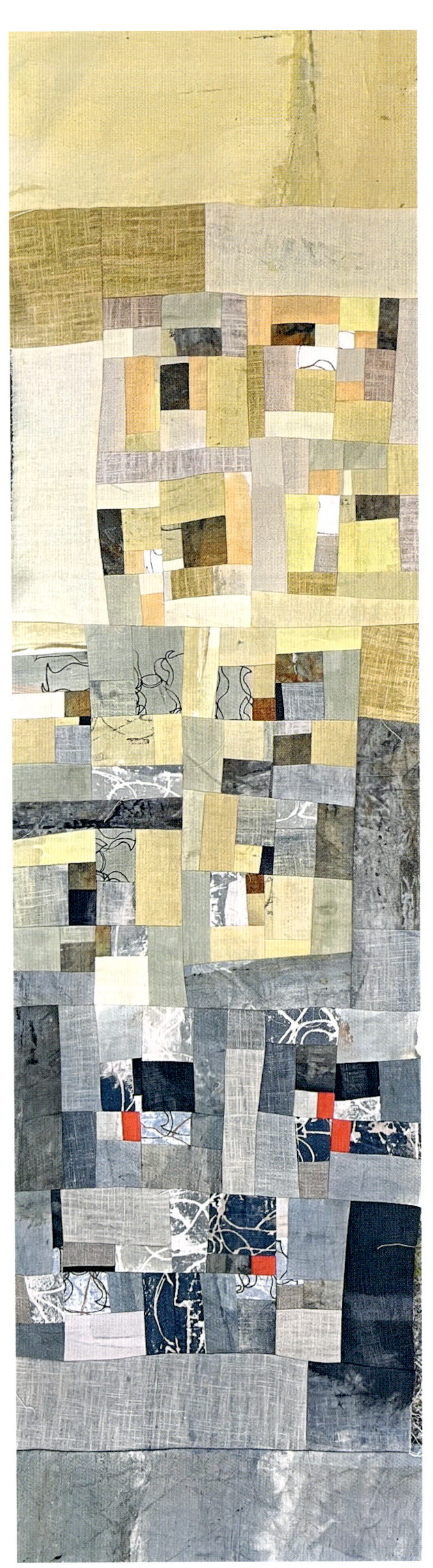

FEATURED ARTIST

Liske Johnson

I worked at Liske Johnson's studio and was struck by her colourful and creative examples of mixed media and patchwork. She tells us more about her work here:

'I run Littleheath Barn Studio in Worcestershire, which gives me a great excuse to play, experiment and meet lots of amazing textile artists. Having specialised in printed textiles while studying textile design, I love nothing more than adding print and marks to cloth. However, recently I have been focusing more on colour than print in an effort to create more space in my work.

'My colour choices and style of work are heavily influenced by landscape and nature and I am always taking photos. My work evolves quite organically. I try not to begin with an end goal in mind; I just allow myself plenty of time to play and explore along the way. It's so important to enjoy the process; if you end up with something you love, it's a bonus!

'In my current practice I use Procion dyes to add colour to natural fabrics, and vintage cottons and linens are my go to. It's worth taking time to mix your colours in order to get the result you are after. I much prefer working with dyes as a paste, so I thicken them with Manutex.

'My favourite way to achieve lovely texture and subtle colour on cloth is with breakdown printing, also known as deconstructed screen printing. This is a process where the thickened Procion dyes are applied to a blank silk screen, allowed to dry, then reactivated by pulling through with more print paste onto soda-soaked fabric.

'When it comes to composition and stitch, I gravitate to an improvised style of patchwork, piecing fabrics together, then often chopping them up further, adding more bits and reassembling them until I am satisfied. I will then add an element of free-motion embroidery along with some hand-stitched details. I love the imperfect line and texture that free machine stitch gives.'

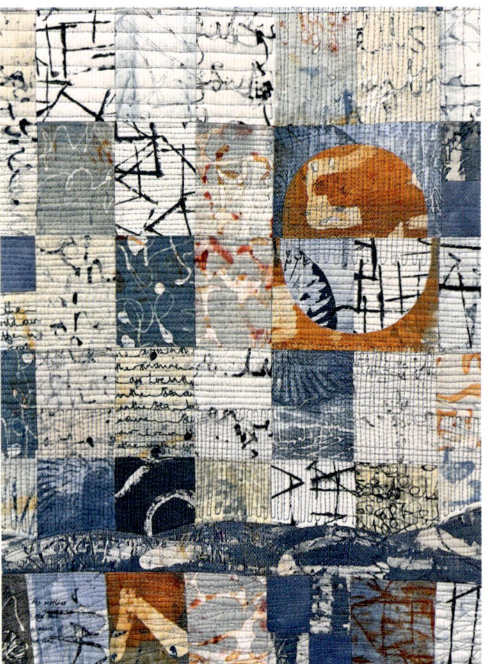

Far left: Liske Johnson, *Untitled*. Hand-dyed improvised patchwork hanging

Left: Liske Johnson, *Live in the Sun*. Hand-dyed and printed textile hanging with free machine quilting

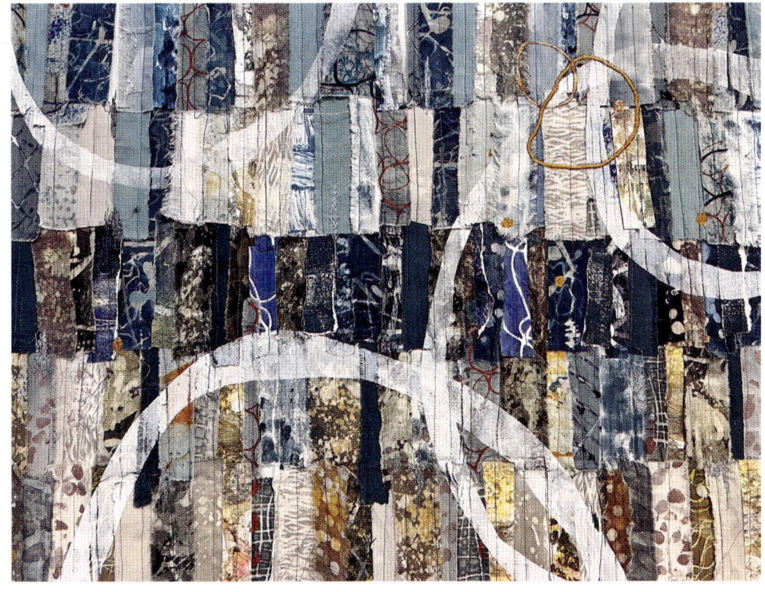

Left: Liske Johnson, *Pebbles*. Mixed media textile collage

Below: Selection of vintage patchwork tops, sourced by the author

ABSTRACT PATCHWORK SHAPES

A good way to get a piece together quickly is to use patchwork, but in a loose and unconventional way. I like to find old patchwork quilt tops to repurpose. They provide a textured, colourful and rich patina to work on. The older and more shredded, the better. I often need to repair them before commencing a project.

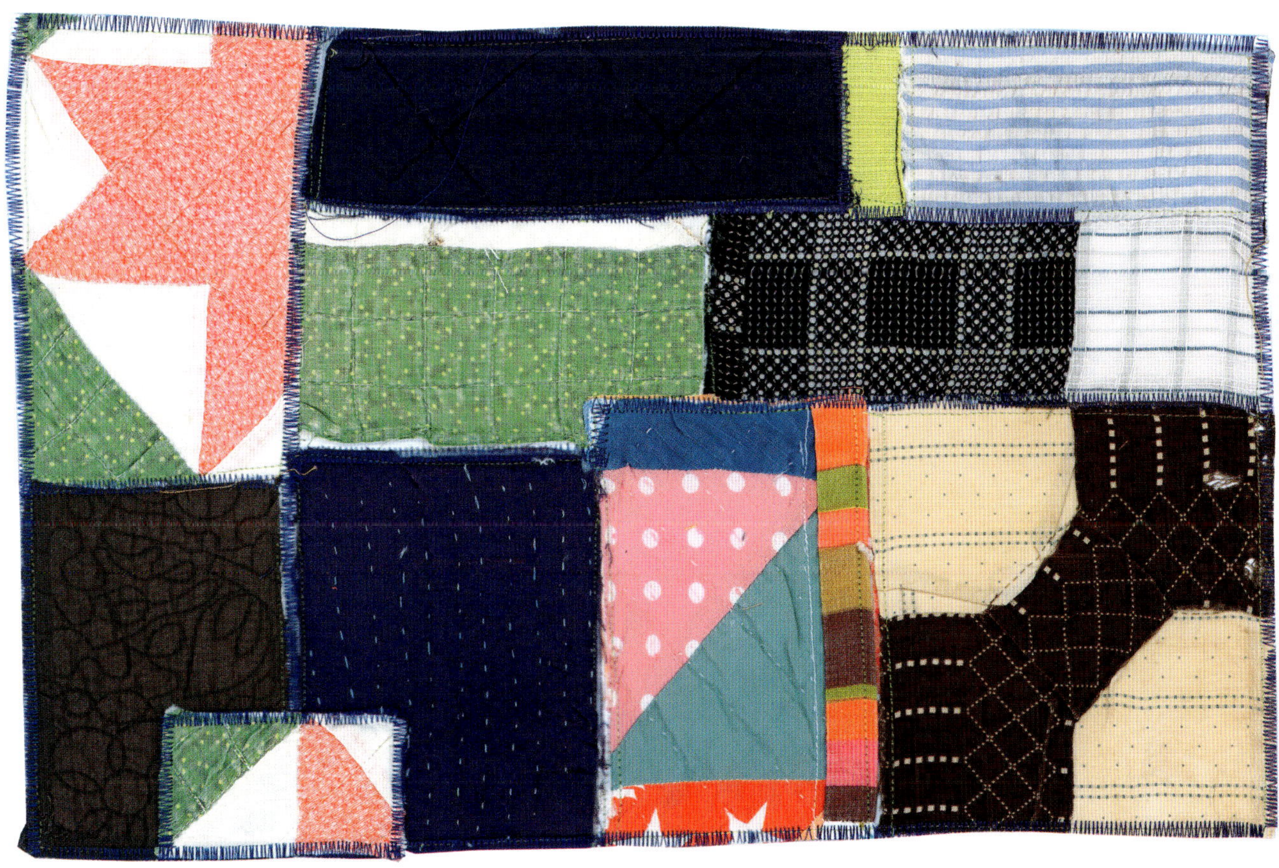

PATCHWORK HOUSE BAG

Using blocks can be an easy way to create an abstract composition. Patchwork and sewing lend themselves to this and in this example I worked on a series of different patches, joining them together and then attaching them to a background. You could use old or incomplete quilt remnants or found pieces to do something similar. Often old quilt fragments of tops can be found in charity shops or online.

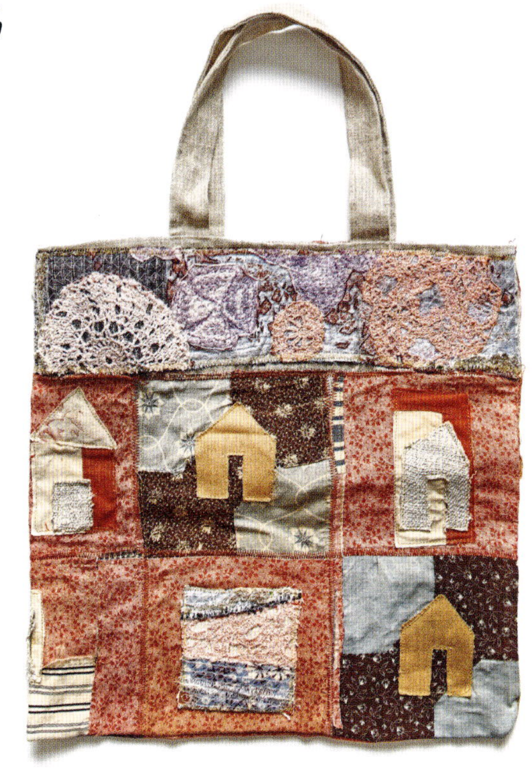

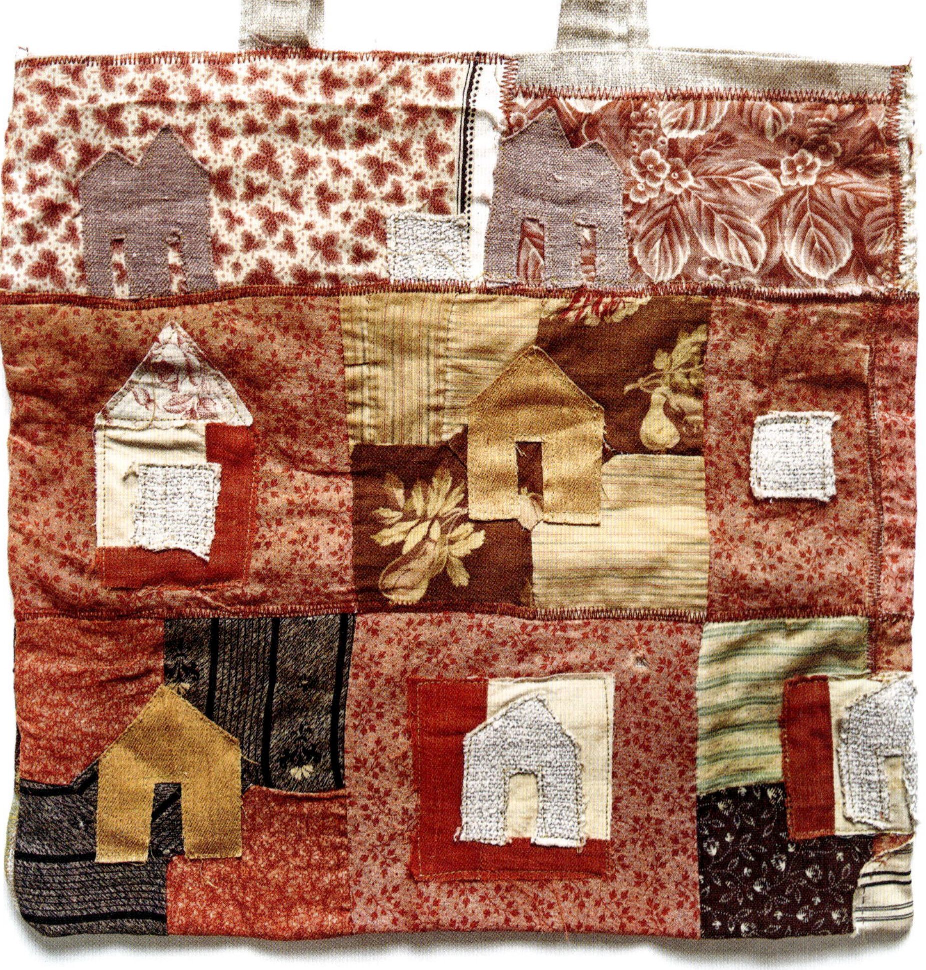

Above: Anne Kelly, *Patchwork House Bag*. Stitched textile

Left: Anne Kelly, *Patchwork House Bag*. Stitched textile

Opposite: Anne Kelly, *Patchwork House Bag* (detail). Stitched textile

WORKSHOP

Making a patchwork house bag

MATERIALS REQUIRED

- A cotton or linen tote bag, opened up to lie flat
- A piece of quilt or quilt top. You can make your own background from scraps of fabric, enough to cover the body of the bag with at least 1cm (½in) seam allowance
- Piece of fabric to line your bag (optional)
- A theme or idea for shapes
- Extra fabric for additional shapes
- Hand- and machine-stitching threads

METHOD

1. Arrange your shapes on the quilt top.
2. Stitch and embellish them onto the background as you wish, by hand and/or machine.
3. Cut your lining to fit if you are using one.
4. Pin the piece together through all two or three layers, ensuring the fabric is even on the sides of the bag. Depending on the weight of the bag, it may be useful to tack (baste) the layers together.
5. When attached, stitch the front and back, inside out and right sides together. Turn right side out. Embellish the handles if desired.

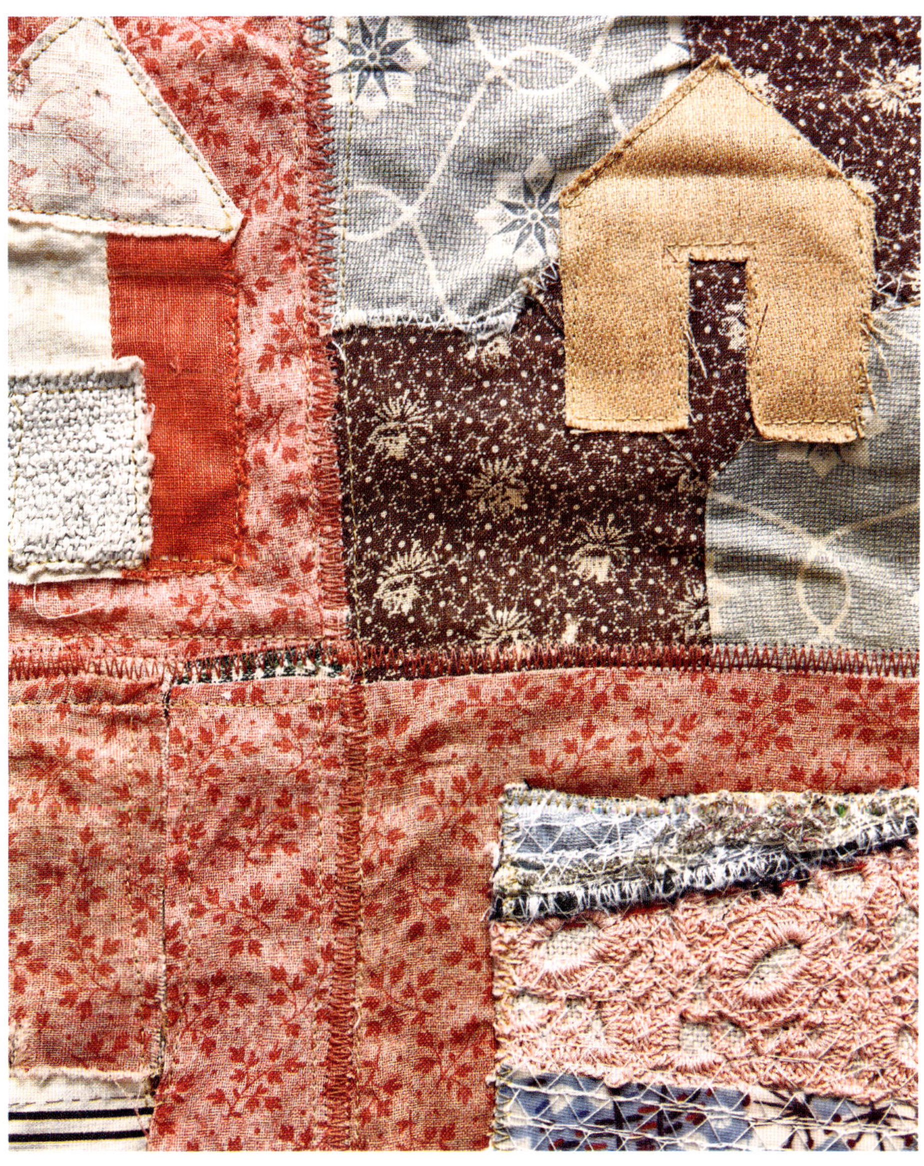

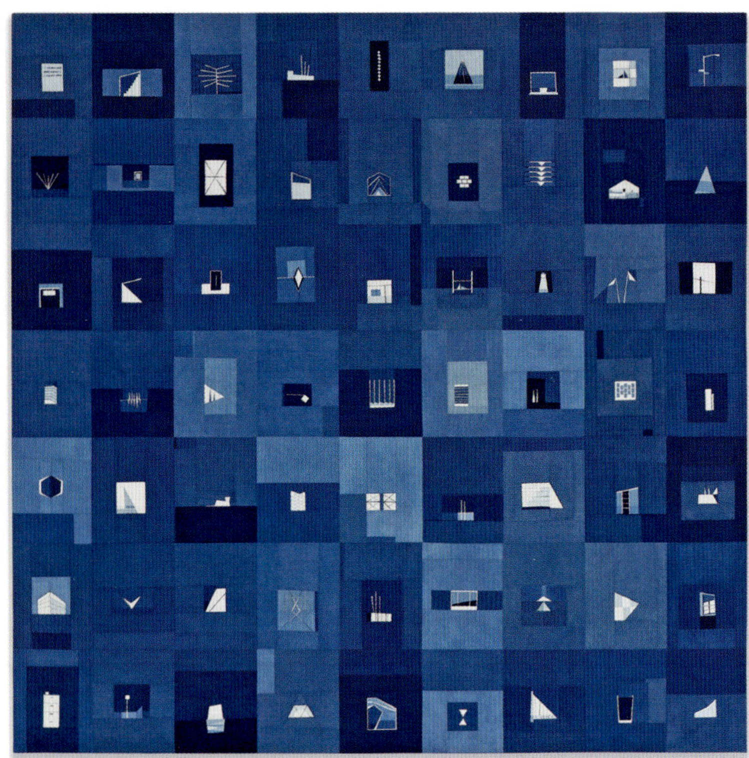

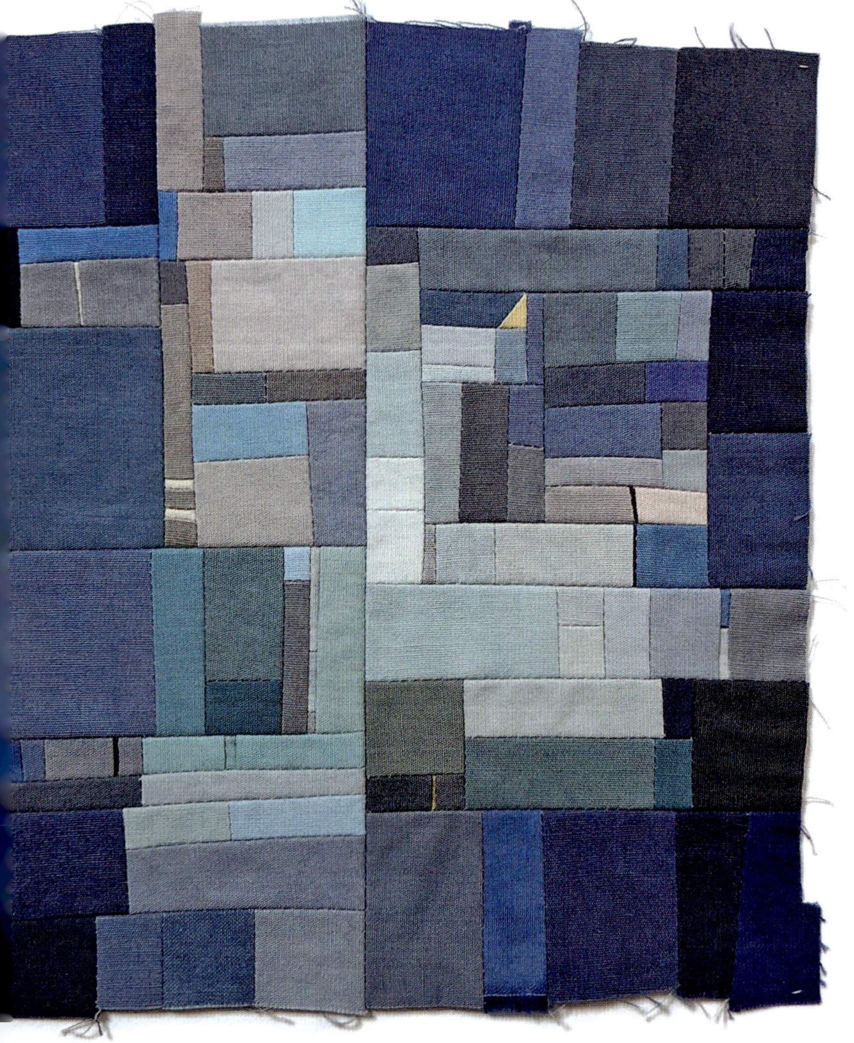

FEATURED ARTIST

Erin Wilson

Erin Wilson works with geometric precision in different scales and explores a variety of colourways, referencing her inspirations. She tells us more:

'I absorb the world in details. I make patchwork with this same micro focus. I like finding how things fit together. Inside the structure of patchwork piecing, I stitch architectural shapes, colour assemblies, and scenes from daily life. I work small, in straight lines and grids. Fabrics are hand-dyed using fibre-reactive dyes and indigo. I don't keep recipes. Each colour is mixed and divided into a gradation, dark to light.

'My current studio is an attic storage room, windowless but private. I sit at my sewing machine with a pile of fabrics on the table and the ironing board to my left.

'For the *Shape Studies*, I work from a sketchbook with simple line drawings. I visualise the construction steps and begin. The process is an intense combination of precision and improvisation. The *Assemblies* are looser, made with the leftovers. They are a warm-up, a cool-down, or, when I am totally stuck, a gentle way to stay in my practice.'

Above left: Erin Wilson, *Shape Study 26*. Quilt

Left: Erin Wilson, *Assembly 18*. Patchwork

Opposite top: Anne Kelly, *Small Houses*. Wood

Opposite centre: Anne Kelly, *Covered Small Houses*. Vintage paper on wood base

Opposite bottom: Anne Kelly, *Covered Small Houses*. Painted canvas on wood base

LITTLE HOUSES

WORKSHOP

Covering a little house

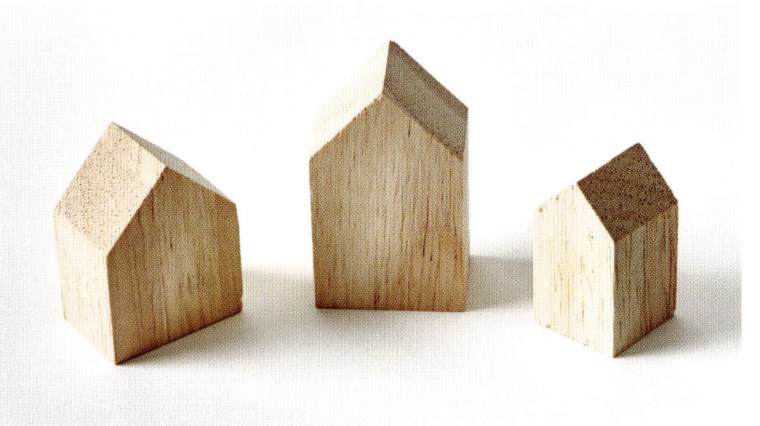

MATERIALS REQUIRED

- Small wooden house shape
- Scraps of fabric and paper
- Felt for base (optional)
- Glue stick
- Scissors, cutting knife
- Varnish, PVA glue

METHOD

1. Draw around each side of your house on paper and/or fabric.
2. Cut paper/fabric to size.
3. Glue in place with glue stick.
4. When finished, coat your piece with varnish or diluted PVA glue to protect it.
5. Cover the base of the house with a piece of felt if desired.

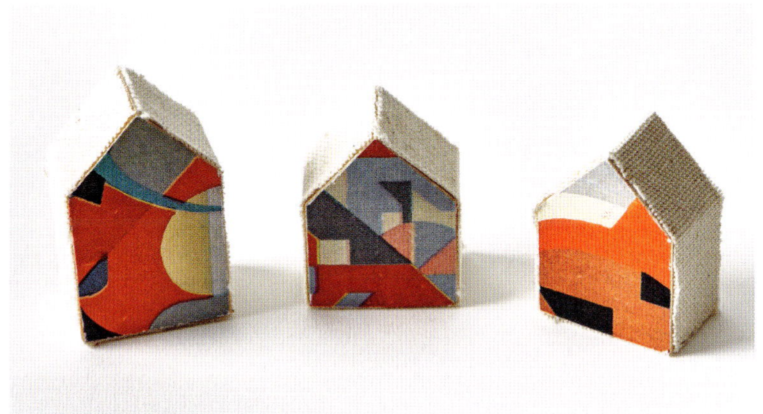

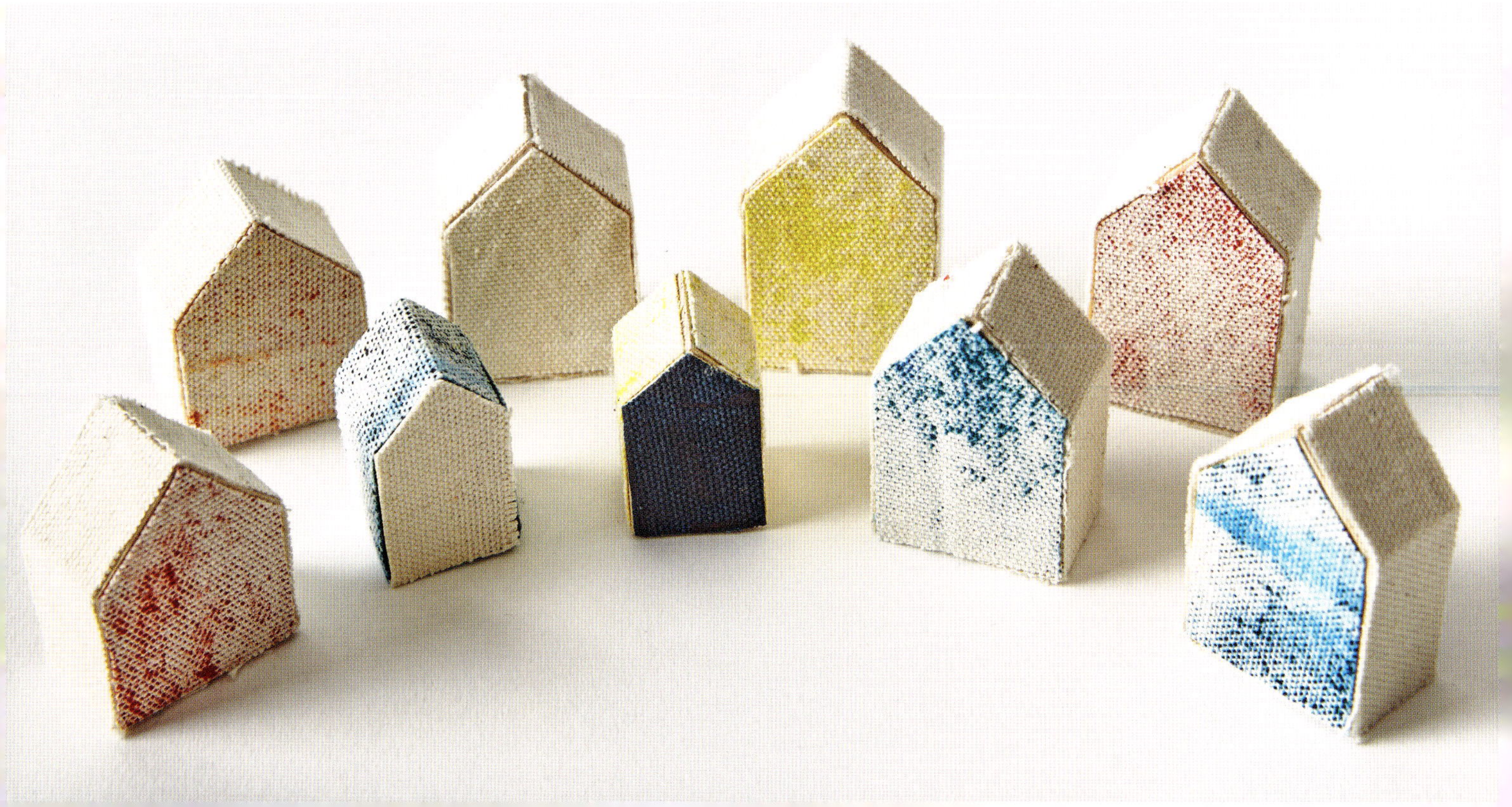

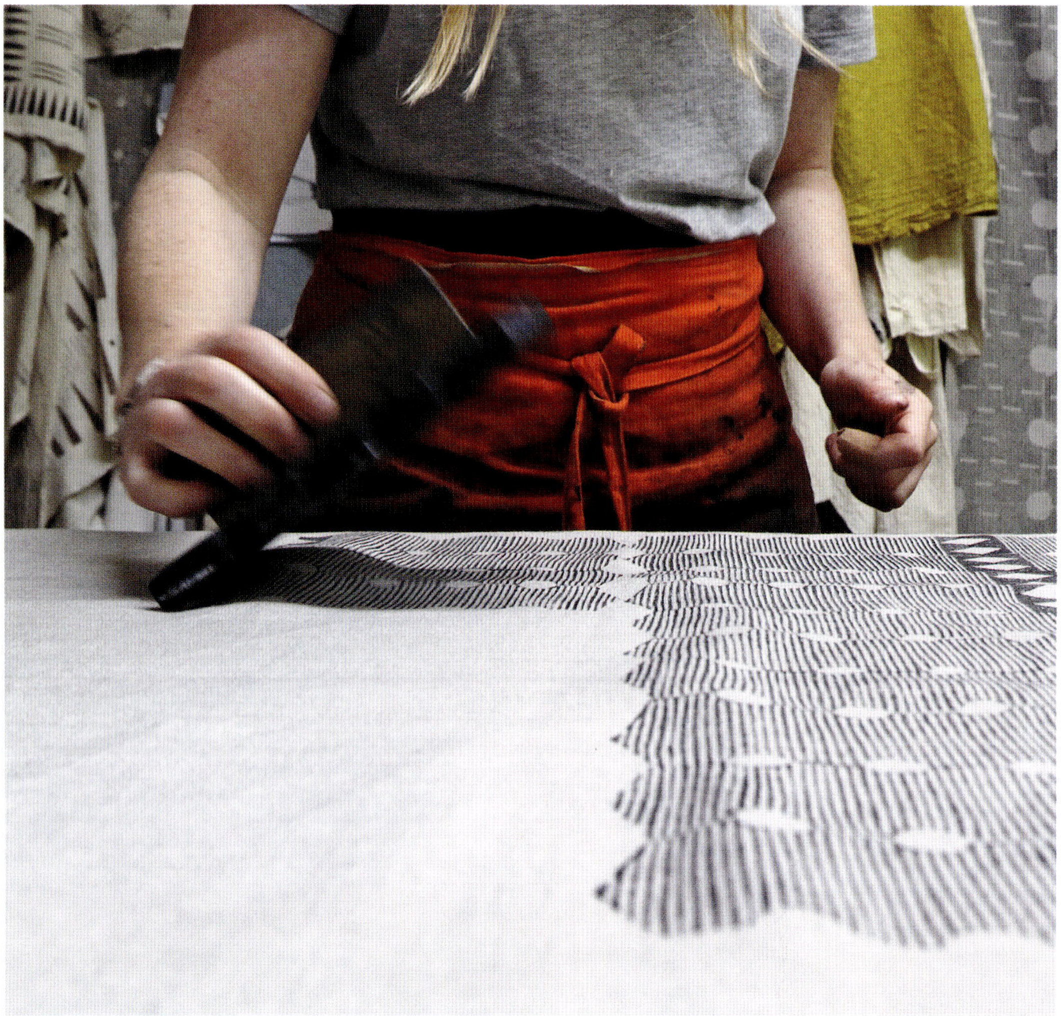

Left: Louisa Loakes, *Laugh & Hail*. Hand block printed textiles

FEATURED ARTIST

Louisa Loakes

Louisa Loakes's work, designed on hand carved blocks, appears deceptively simple and is rich in texture. It is reminiscent of some of the great women printers of the 1930s, as she explains:

'I began my journey with the block after a trip to India, where I first discovered Indian textiles. From here I began to explore the craft of traditional block printing. I found a connection; a language between the process and my abstract paintings and charcoal line drawings; through pattern, bringing delicate lines and simple geometric forms together. The hand block-printing process produces irregularities; these honest but sometimes imperfect marks are what I love about a piece of block-printed fabric. As the block twists and turns across a length of fabric, a rhythm runs through it, giving it energy and life.

'I regard a length of printed fabric like a canvas or a painting. Feelings, intuition, play and process are all central when creating my work. I work freely by cutting straight into the lino like a sketch; this sketch usually ends up being my final piece/block. It is important that I hand-carve my own blocks: each mark I make is my own and honest.

'The printing process is very physical: my eye is my guide that speaks to my hand, and they rely on each other to establish a meditative rhythm when printing. I print using a monochromic palette. After, I hand-paint a dash of colour, usually burnt orange. Adding the hand-painted colour brings out another energy in the pattern/work; another pulse or spirit that runs through the length of fabric. When I stand back and look at a finished length of printed and painted fabric, I see a finished canvas.

'Running alongside traditional textiles in the world, I look to the work of many artists: Agnes Martin, Sonia Delaunay, Matisse, Picasso, and artist block printers such as Enid Marx and Barron & Larcher.'

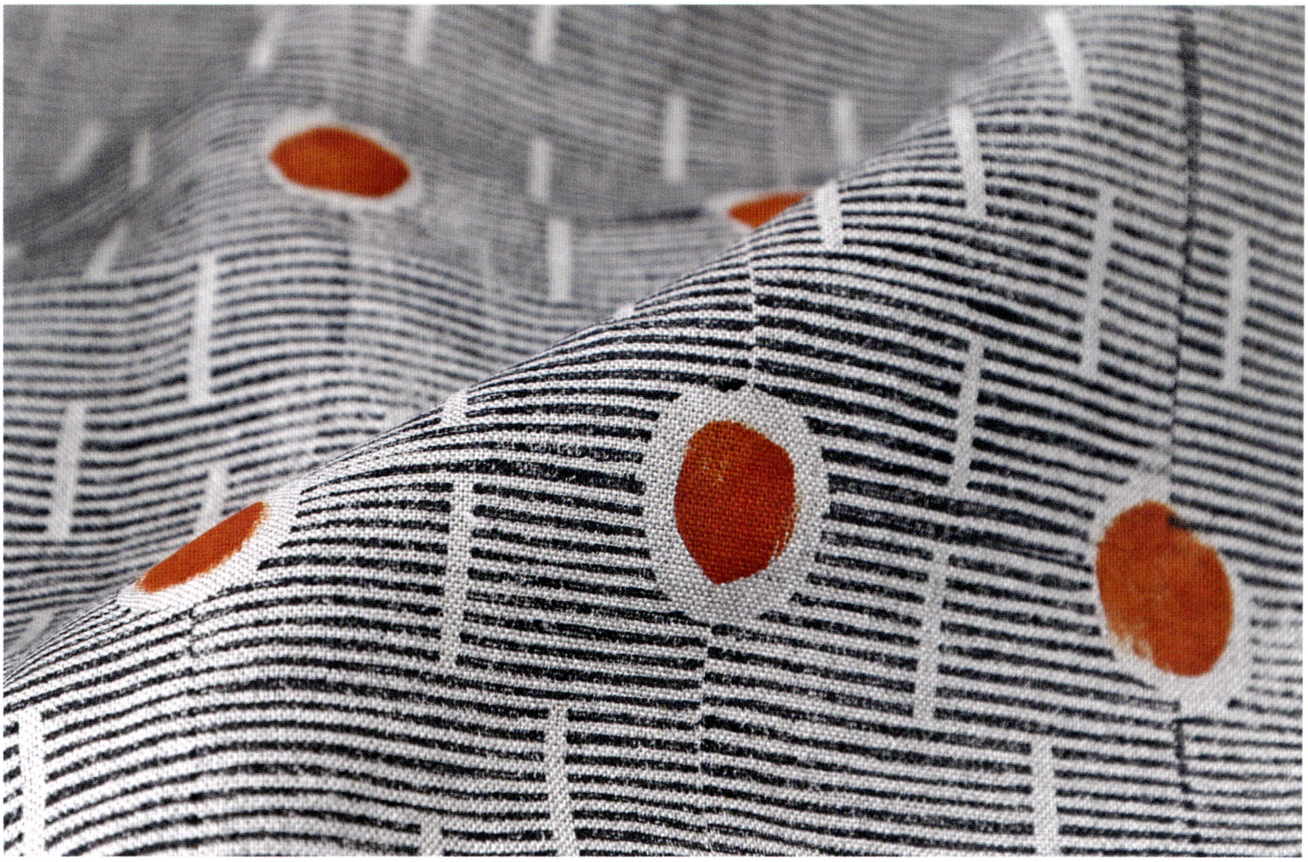

Above: Louisa Loakes, *Saffron Spot*. Hand block printed textiles

Right: Enid Marx, *Chevron*. Moquette sample

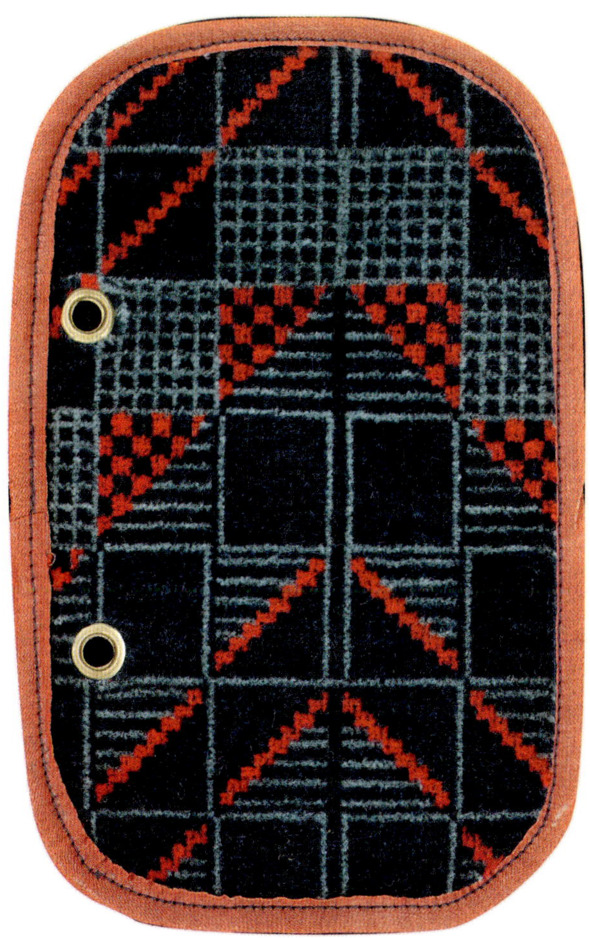

FEATURED ARTIST

Enid Marx

Enid Marx (1902–1998) was a multi-talented artist and designer, renowned for her designs for utility furniture and the London transport system. She and her partner, the historian Margaret Lambert, were passionate about folk art and displayed their designs at Compton Verney; their collection is a now a permanent collection on display at Compton Verney museum.

WORKSHOP

Rubber stamp prints

MATERIALS REQUIRED

- Rubber erasers or pieces (available online)
- Cardboard for mounting shapes, PVA glue
- Small sponges and acrylic paint for printing
- Cotton or linen fabric for printing
- Plastic to protect surfaces
- Cutting board
- Cutting knife
- Pencil and paper for designs, tracing paper

METHOD

1. Make simple shapes for your designs – I used sea-inspired shapes from an old tin.
2. Cut out your shapes using a knife or heavy scissors.
3. Glue onto card pieces and leave to dry.
4. Using small sponges, gently brush the surface of your block with a thin, even coat of acrylic paint.
5. Gently press the block onto your fabric and repeat to create a pattern.

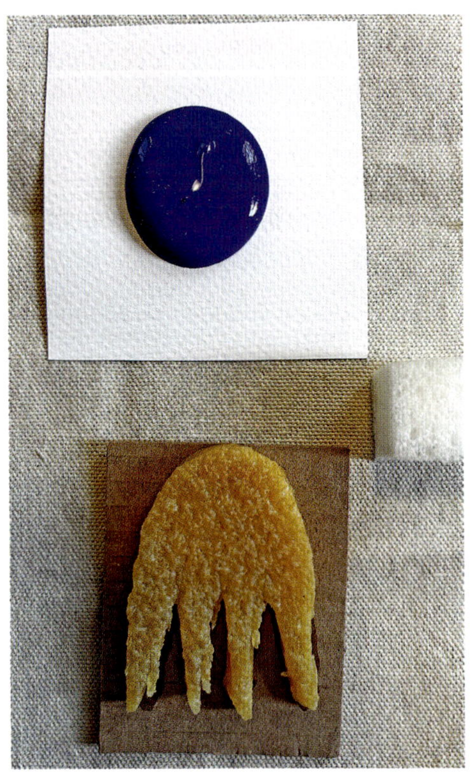

Left: Anne Kelly, printing block and acrylic paint

Right: Anne Kelly, *Sea Shape Prints*. Acrylic print on canvas

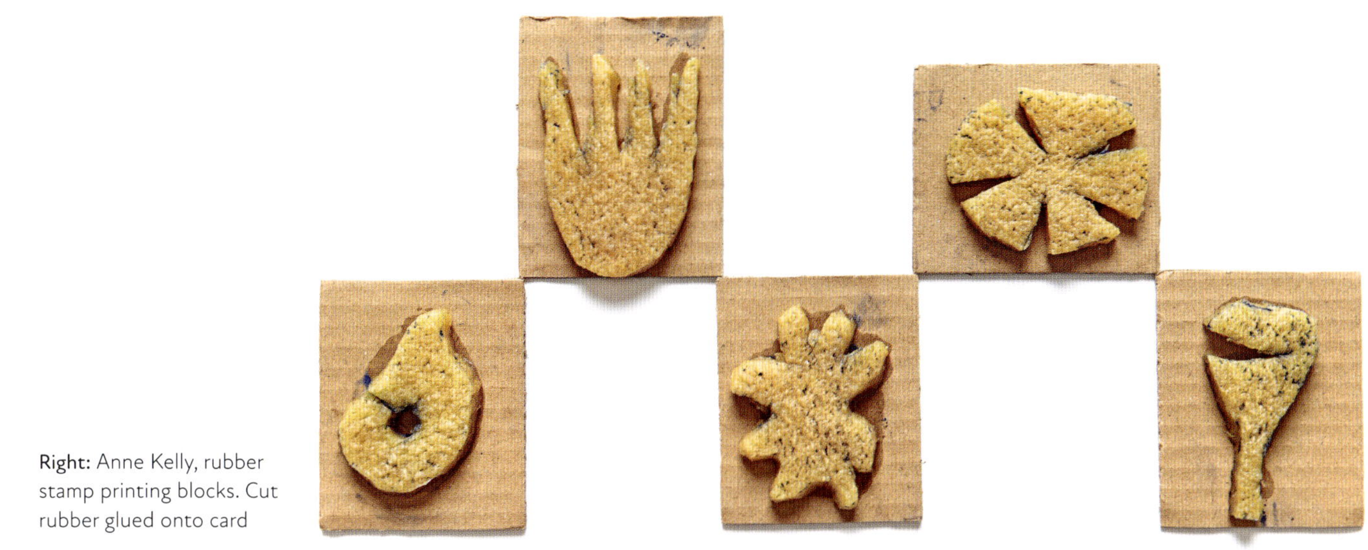

Right: Anne Kelly, rubber stamp printing blocks. Cut rubber glued onto card

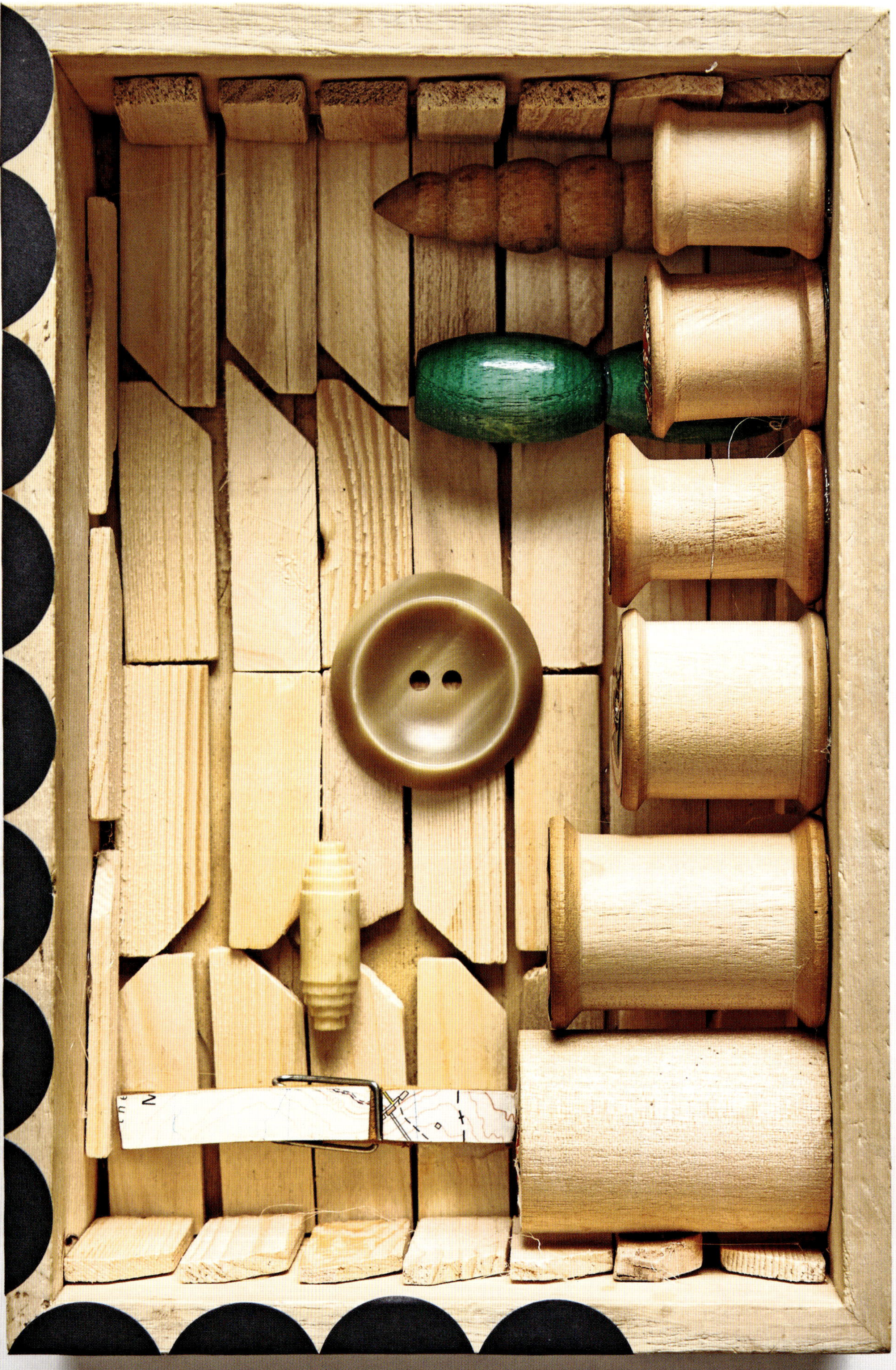

Chapter 3

SURFACE PATTERN AND MIXED MEDIA

'There is no abstract art. You must always start with something. Afterward you can remove all traces of reality.'

Pablo Picasso (1881–1973)

Opposite: Anne Kelly, *Inspiration Box, Shape and Pattern*. Mixed media in wooden box

In the transition from figurative work to abstract, it is helpful to use elements and techniques that enable expression and variation in your practice. Working with surface pattern and mixed media can be helpful. Often, working within these disciplines can open up other possibilities.

In this chapter we look at methods that promote pattern and mixed media. By pushing ourselves outside of our comfort zone we can discover new and exciting opportunities and outcomes. We will be looking at diverse techniques including digital print, textile and paper collage, screen printing and stencils.

Below: Anne Kelly, *Abstract Collage*. Mixed media textile

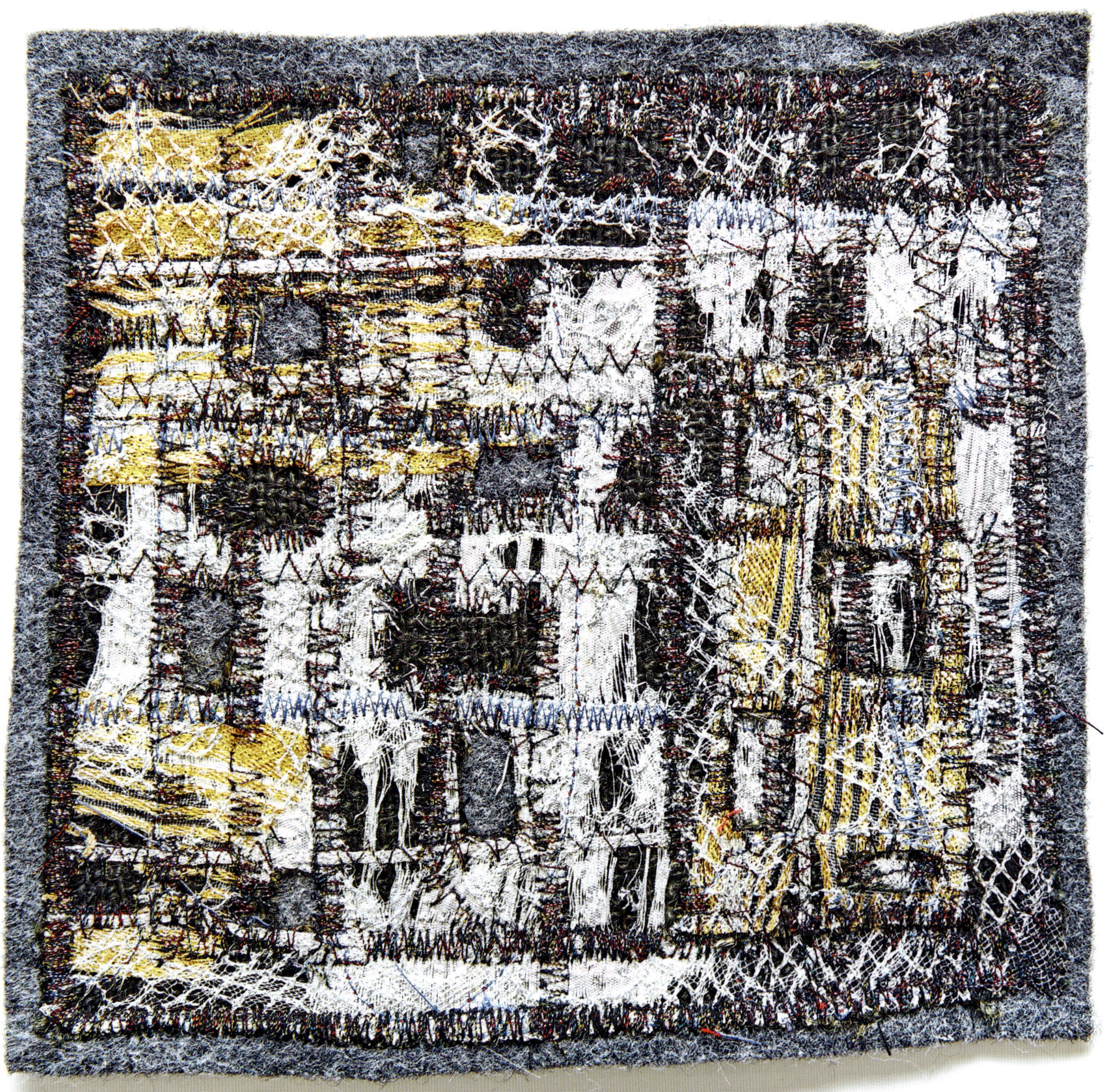

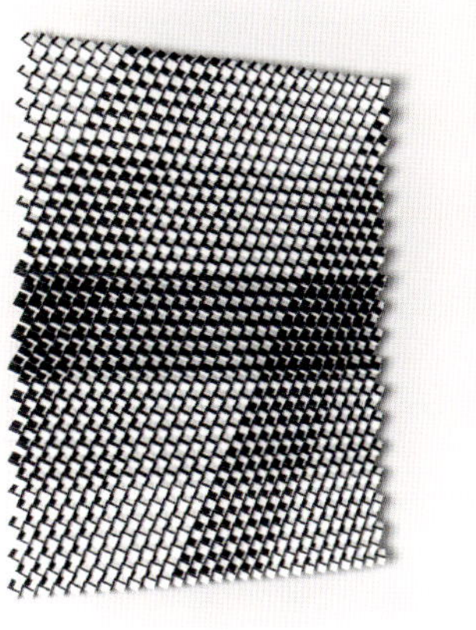

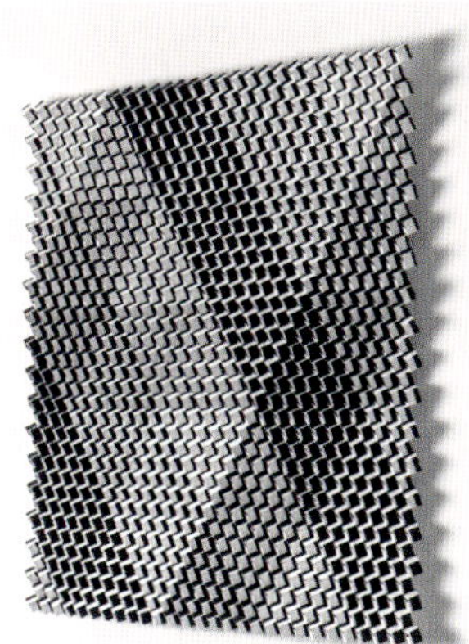

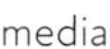

Above: Dail Behennah, *Shift*. Mixed media, three-directional plaiting

Below: Dail Behennah, *Resonance*. Mixed media

Bottom: Dail Behennah, *Resonance* (detail). Mixed media

FEATURED ARTIST

Dail Behennah

Dail's work is visually arresting and belies much experimentation and a complex understanding of shapes and their juxtaposition. She speaks about her work:

'I am currently working almost exclusively with paper to make undulating surfaces using a three-directional, three-dimensional plaiting technique. These works may be very large and free-hanging, or smaller and framed. Light and shadows play across the complex surface, further emphasised with graphite or ink, and different facets come into and out of focus as the viewer shifts. I enjoy investigating the wealth of complex geometry to be discovered in the structure.

'*Shift* was the first work in which I marked the paper strips with Indian ink. I drew a plan of the pattern before I started making, but once completed I was surprised to find how different the work looked as one moved past it, different facets of each block coming into and out of sight.

'*Resonance* is a completely different kind of work. Guitar strings need to be changed regularly and I have accumulated many over the years. The stiff and springy nature of the material dictated the coiling technique. The strings have been stitched together to make a large disc, with thinner strings in the centre and thicker ones towards the edge. The wavering lines of stitching reveal the work of the hand, and there is more air than material, which renders it transparent. The brass ball ends create a pattern that contributes to the strong sense of circular motion. As the strings are all the same length, the pattern is determined by the gradual increase of the circumference and is not planned. This piece contains music and time, and took more than eighty hours to stitch.

'When I work, the slow process and manual concentration free my mind to think. An idea and the making of it stimulate and change each other. By the time I have finished one piece I have imagined another ten, of which only one will be made. I work with care and precision, creating well-finished pieces that will last for many years.

'My work has developed and changed, but my preoccupation with line, light and shadow has remained constant. Above all, I try to give a feeling of calm, but not stillness.'

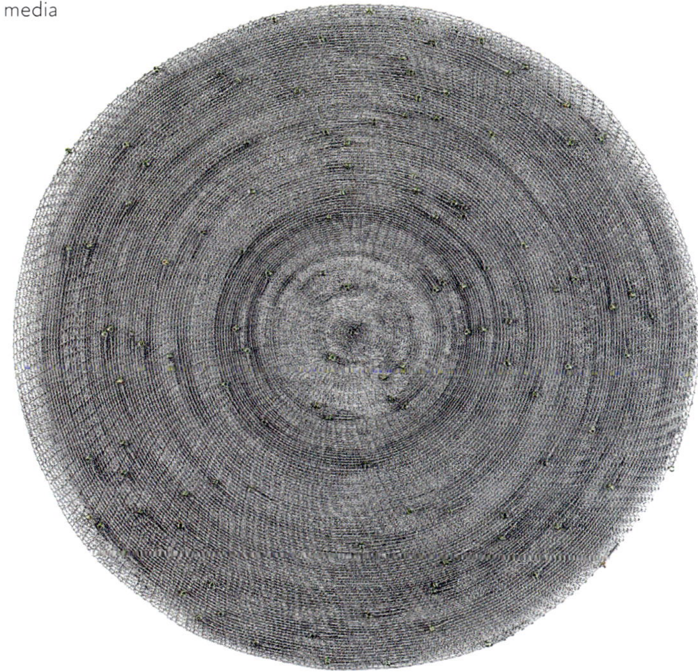

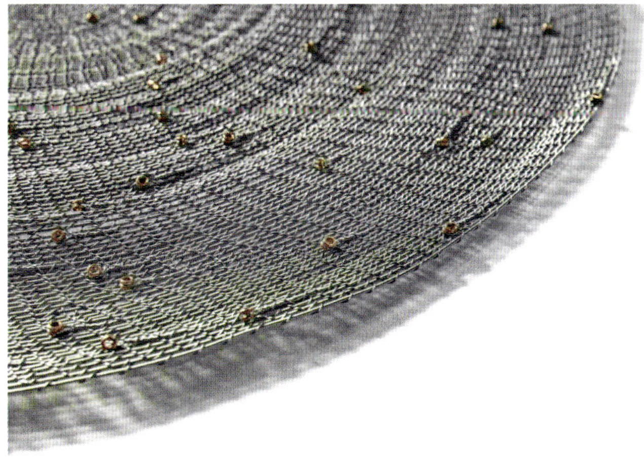

'JUNK' COLLAGE

This is a fun exercise to create a mixed media sample that can be used as a starting point for a body of work or presented as a work of art in itself. Have a look through drawers, toolboxes or sheds where you keep old ephemera, or anywhere you might find interesting shapes, textures and patterns.

WORKSHOP

Making a junk collage

MATERIALS REQUIRED

- Card, wood or discarded book cover for background
- Main elements, fabric scraps, wood pieces, buttons, metal ephemera
- Paint – acrylic and/or fabric
- Wool, embroidery thread, ribbon, string
- Glue – stick, gun, PVA
- Varnish to finish, if desired

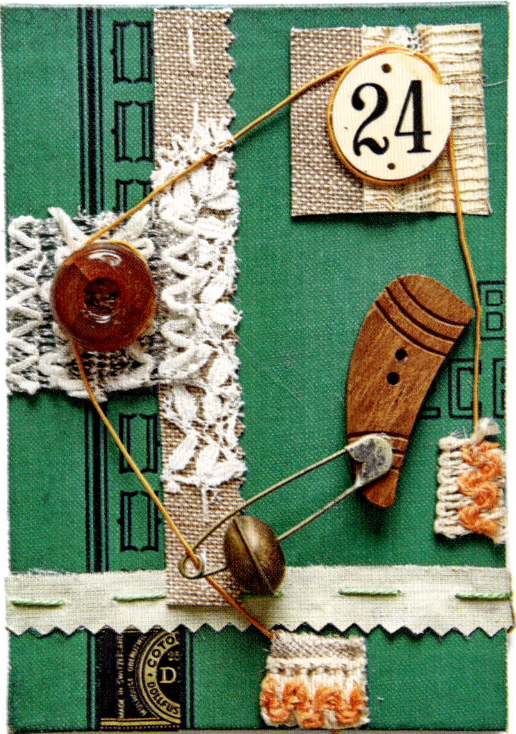

There are many possibilities with this project. I suggest you choose your materials in stages, as it is an important part of the process:

- **A base** – a piece of wood, card or even an old book cover. The more distressed and interesting the surface, the better. It will provide a great backdrop for everything that comes on top.
- **Elements** – these can be found objects (scraps of wood, plastic, textile) or fabric remnants (any fabric type, shape or texture).
- **Linear marking** – string, ribbon, wire, wool.

METHOD

1. Start with your base(s). Look at the shape. Does it need trimming or adding to? It is feasible to work on more than one little piece at a time; indeed, this project lends itself to working on a series. You may want to add colour at this stage.
2. Play with your collections of objects on top of your base. Move the elements around, taking care not to cover the background completely. When you are happy with the arrangement(s), take a photo to check and then glue them down onto the base.
3. Add stitched/linear marks where appropriate. These will add detail and emphasise the shape of your elements. It may feel right to add more painted areas at this stage.
4. Finally, you can varnish your piece(s) if desired.

Left: Anne Kelly, *Junk Collage 1*. Mixed media on card

Opposite: Anne Kelly, *Junk Collage 2*. Mixed media on card

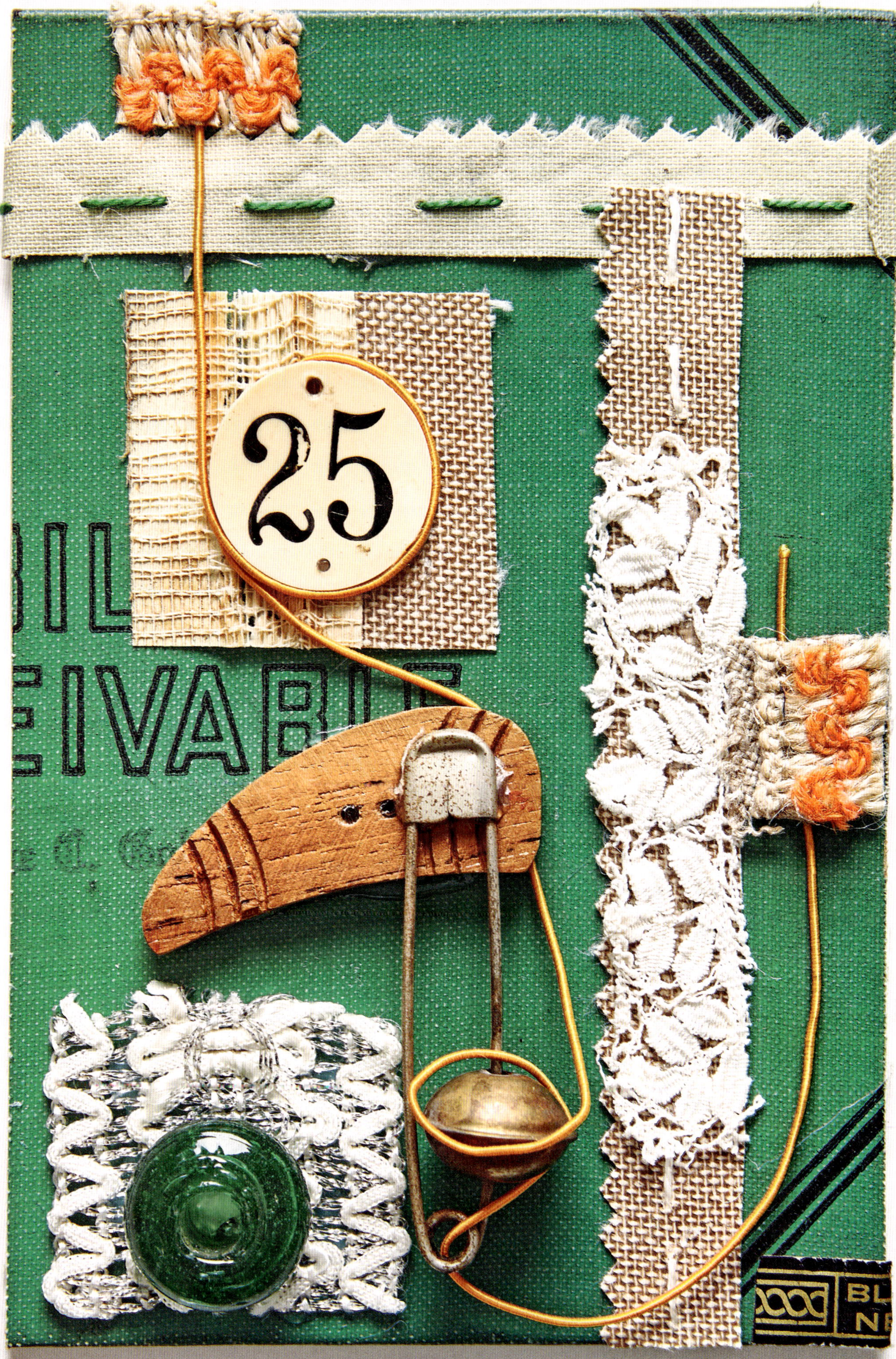
25

FEATURED ARTIST

Louise Baldwin

I have long admired Louise Baldwin's work and her transition from more figurative pieces to abstract collages. She tells us more here:

'There has always been an element of abstraction in my work, a mark that describes a movement or sound, layered images that blur boundaries placed together with something more literally observed like a plant form or a portrait. In this series, *Temporary Condition*, I would say that observation still lies behind the abstracted images, but perhaps on a less obvious and more emotional level.

'I started working on this series in response to the physical and emotional impact of building a new house and moving from an old home. Clearing out my studio I found all sort of bits and pieces that I couldn't part with, and started to push them together in stitched assemblages. I worked directly with the materials, seeing how they connected with each other, pretty instinctively to begin with. I was recording the house build at the time, with scaffolding and wiring, noises, gaps and things tied down, and this was reflected in how I put these mixed media pieces together.

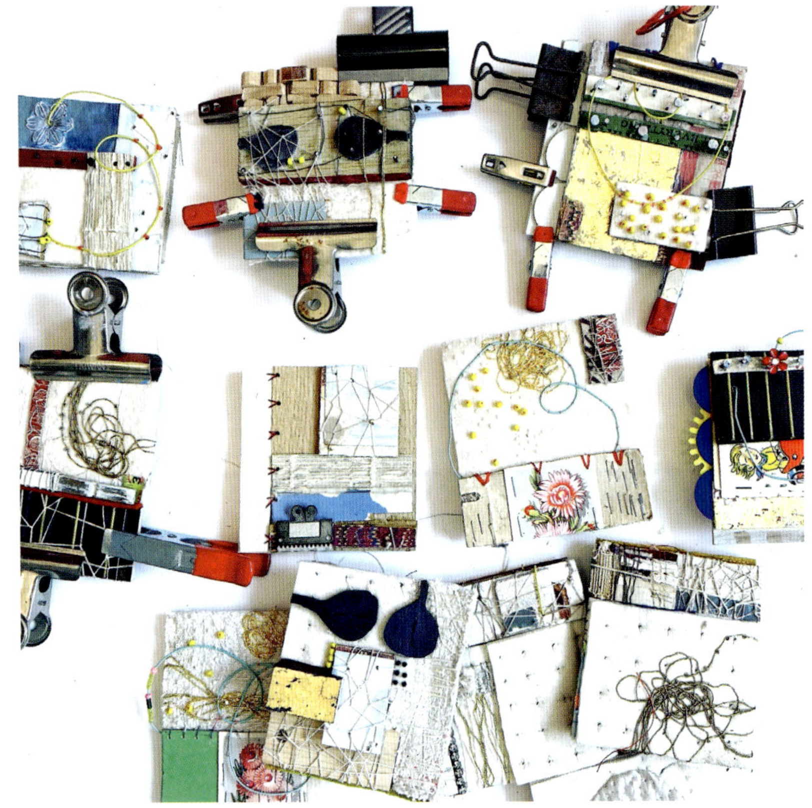

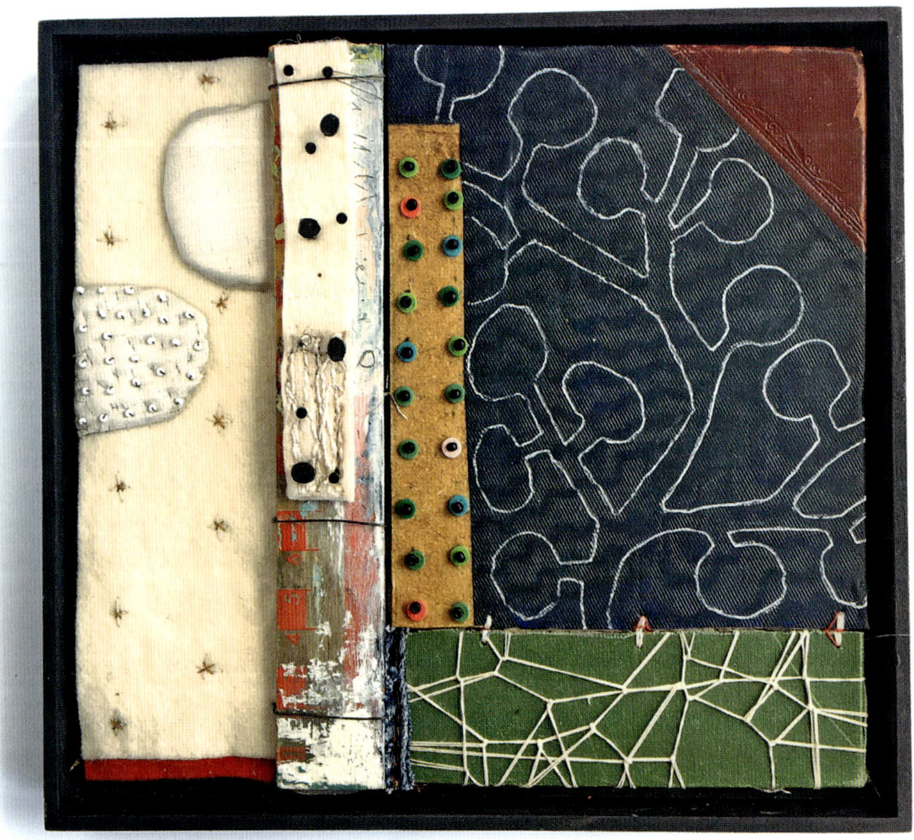

Above: Louise Baldwin, *Under Construction* series. Mixed media

Left: Louise Baldwin, *Temporary Condition*. Mixed media

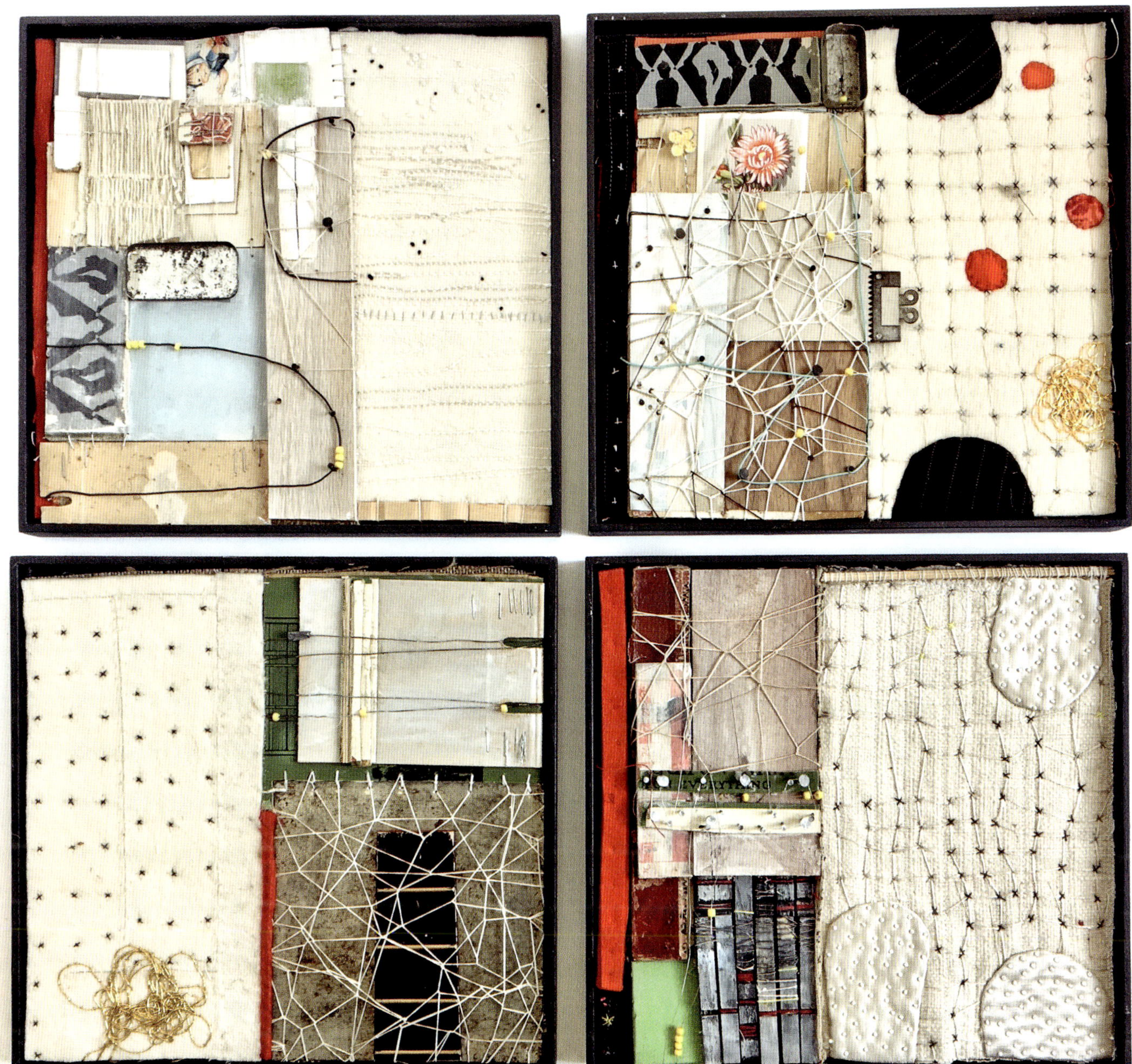

'I used a thick wool felt as a contrasting material, taken from the chaotic and worn materials I discovered in the depths of my old studio. Its intention is to be insulating and calm, functional and almost a blank canvas. The materials are held together with staples, binding, hand stitch and random weave stitch. It appears to be a bodging type of approach but in actual fact it's a very slow process. It involves squeezing things together and shifting them around so that they work.

'Stitch is always used to hold, draw, overlay and evoke ideas. The materials and scale change, the images and drive change, but the stitch remains key.'

Above: Louise Baldwin, *Temporary Condition* series. Mixed media

TEXTILE AND PAPER

Look through your collections for interesting samples of paper and scraps of textiles that lend themselves to collage. Avoid any strongly patterned items (unless abstract, of course) and create a storage box or folder for them. That way you can dip into them as you need them for your work.

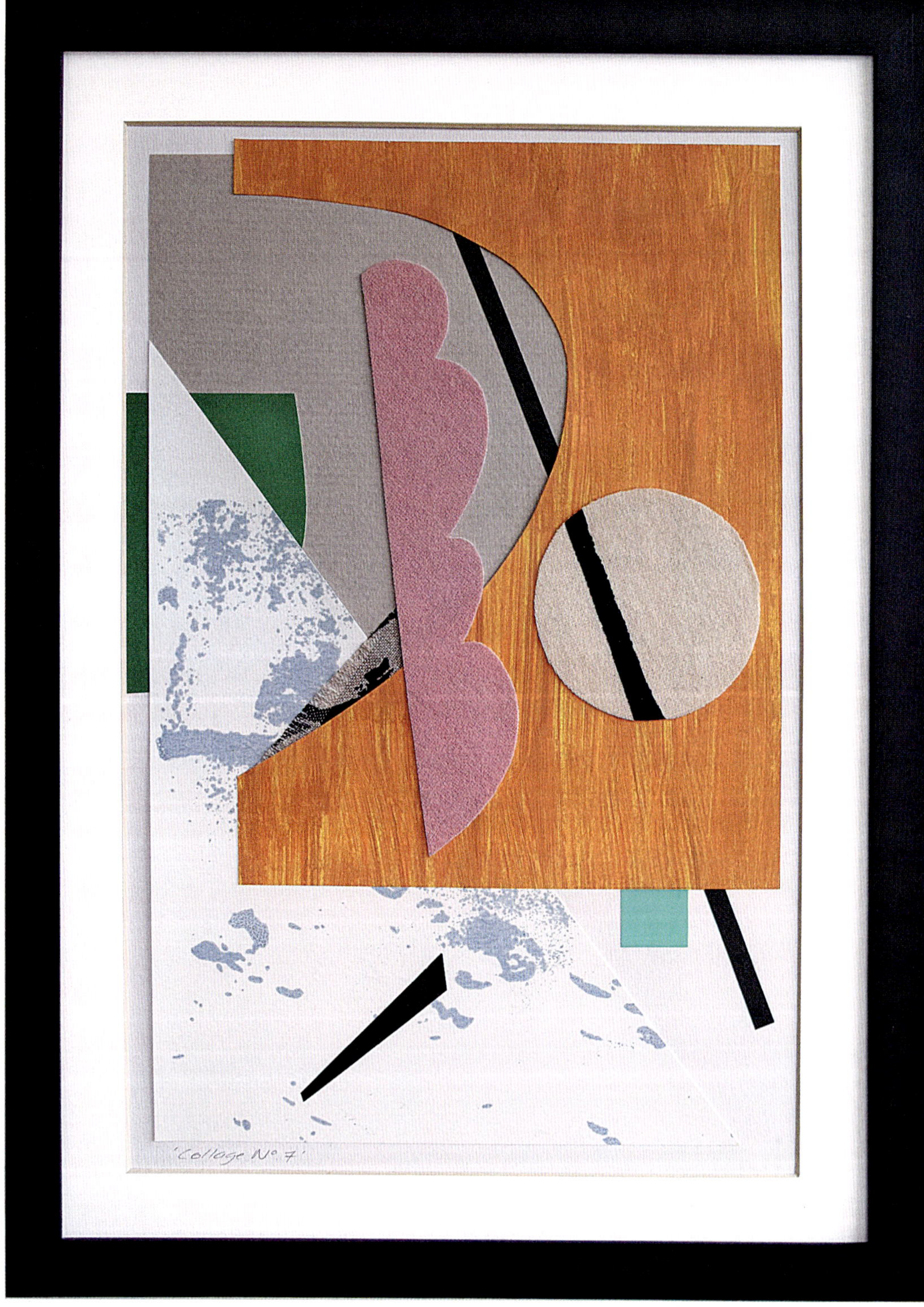

Opposite left:
Michelle House, *Hoop*. Printed textile
Bench: Mandie Beuzeval

Opposite right:
Michelle House, *British Academy Commission*. Hand-printed wall hangings

Left: Michelle House, *Collage No. 7*. Paper and textile collage

FEATURED ARTIST

Michelle House

Michelle House uses collage innovatively, with a strong colour sense and shapes that interact in a striking way. She explains her practice here:

'I fell in love with print during my textile degree course in London. Alongside print, I also created large paintings that had elements of collage in them, and spent many hours in the darkroom.

'My mother always had a sewing project on the go, which sparked my love of fabric. The tactile quality of velvets, corduroy, and satins along with the vibrant patterns of the 1970s made a lasting impression. Scraps of fabric were kept for use another time, along with zips, buttons and ribbons – this is a habit that I inherited. I keep offcuts from my printed and painted textile artworks, along with samples of print and differently textured fabrics. These range from silk, linen and wool to the denim rescued from jeans that have been cut down to shorts.

'From this "paintbox" of fabrics I find inspiration: the edges of painted colour, a printed line here or there, a frayed edge or coarse texture, waiting to be incorporated into a new abstract composition. I approach a collage in a similar way as I would a textile print: there is usually an area of printed pattern, flat colour, texture, with a balanced composition. My paper and textile collages often have an area of relief too, a bold geometric shape overlapping a printed paper and perhaps punctuated by an area of vibrant colour. I love the juxtaposition of the different surfaces, how they absorb and reflect light differently, their textures adding to the artwork in the same way a smudge of charcoal or a brush of watery paint would to a drawing using more conventional art materials.

'Pattern, and the forms found within architecture, have featured in my work for many years. Since moving to London in the late 1980s, its ever-changing architecture and urban environment have constantly inspired me. Through photography, print and my love of colour, my style has evolved over nearly thirty years. Collage has always been part of my work: whether in those first large paintings, studies in my sketchbooks, overlaying print onto textiles, or piecing fabrics together as textile artworks, it's a medium that is integral to my work.'

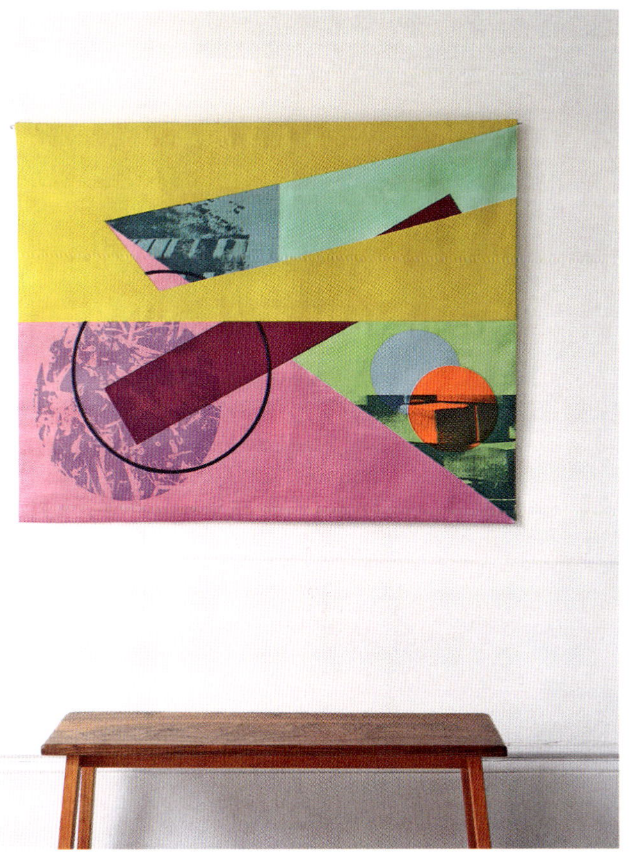

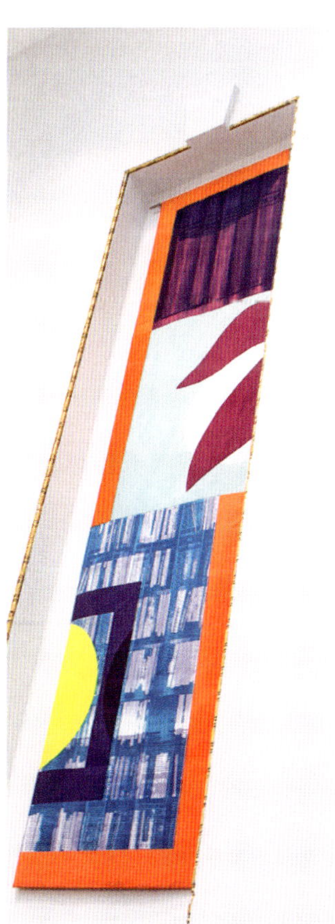

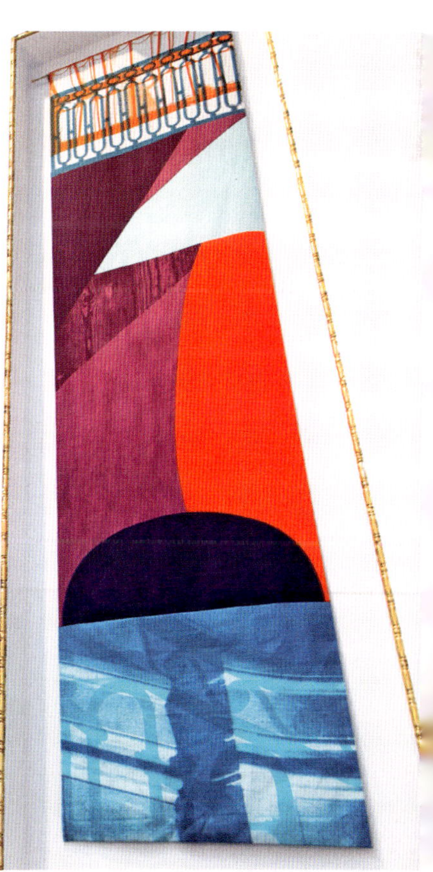

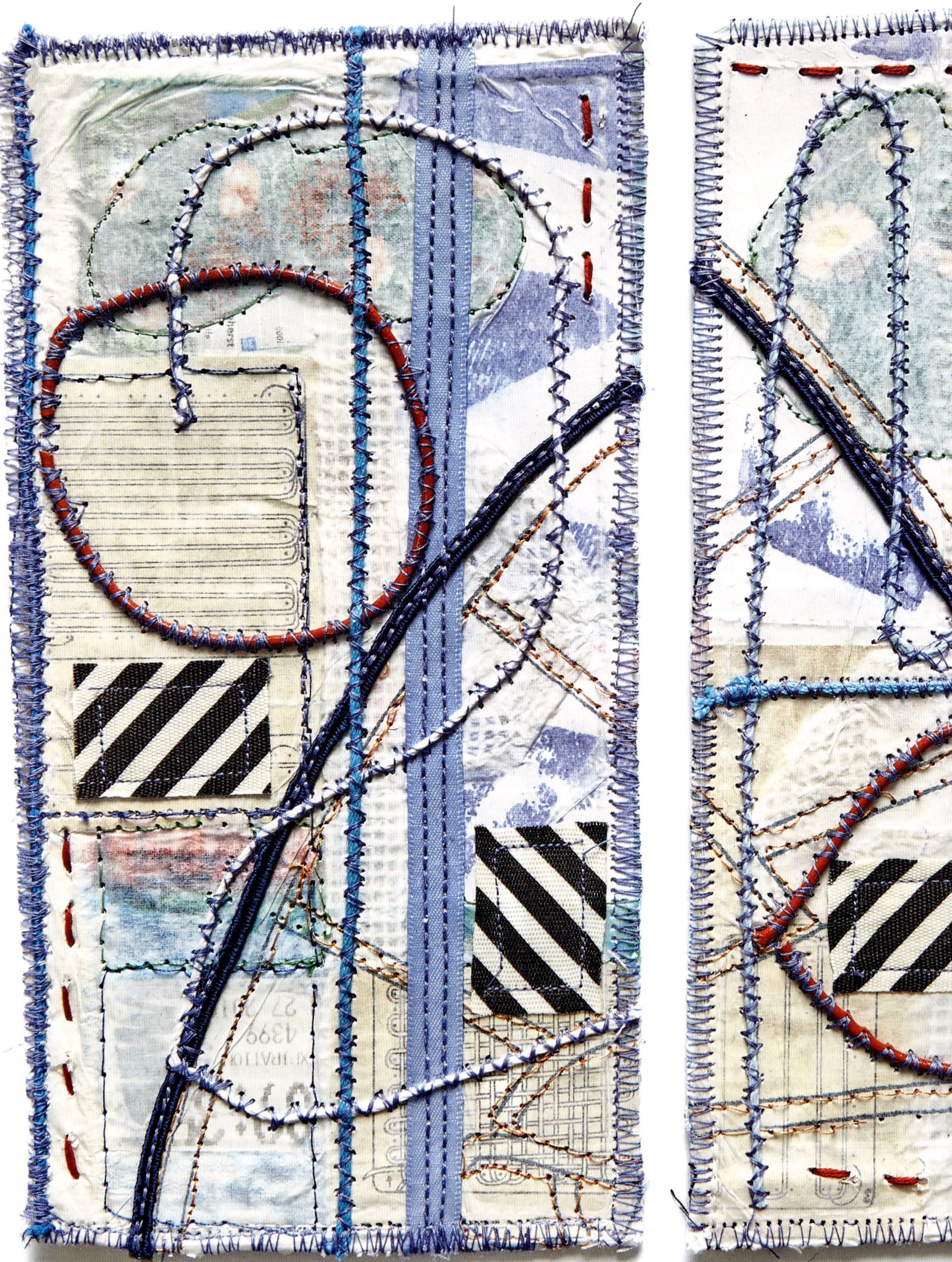

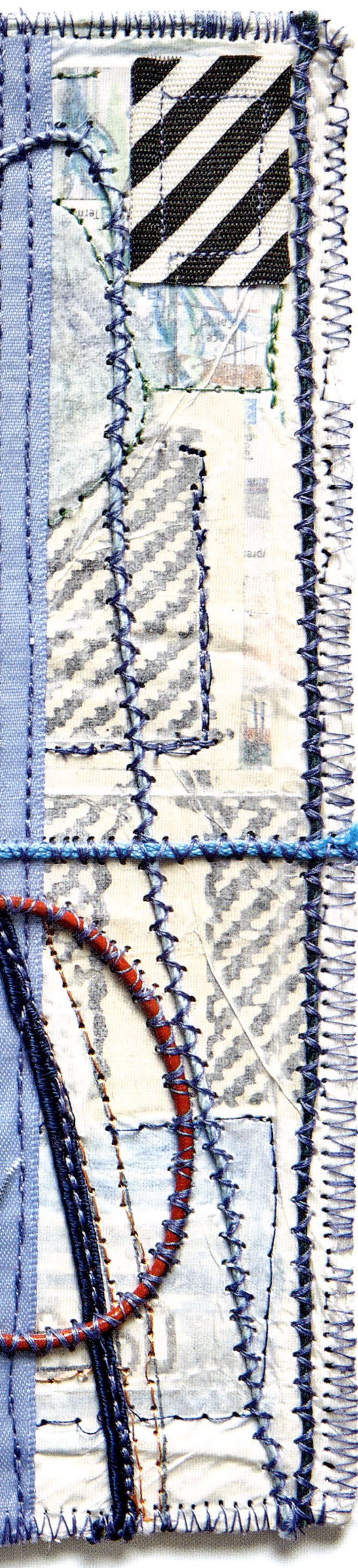

Left: Anne Kelly,
Paper Collages 1 and 2.
Paper and stitch on card

WORKSHOP

Making a paper collage

Why include paper collage here? Working in paper can be very useful for exploring ideas, compositions and colourways before committing to cloth. In this chapter and in Chapter Six, there are suggestions and examples of paper and paper/cloth collages. This is a good project to explore as a series.

MATERIALS REQUIRED

- Paper or thin card for background
- Paper scraps of different varieties (tissue, crepe, newsprint, cartridge, wrapping, book pages, shopping bags, thin card, textured card, laminated paper/card). Your recycling box may contain some interesting items
- Paint, coloured pencils, markers as desired
- Glue stick, PVA glue (diluted) for varnishing
- Scissors, scalpel, cutting board

METHOD

1. Start with a base. Look for a striking or subtle surface as desired.
2. Tear or cut the main pieces you wish to use over the top (a torn edge can become a feature in your design). Play with the arrangement. Photograph to check before gluing down.
3. Add lines with media as desired to create movement in your piece(s).
4. Varnish your piece – make sure that all media used are waterproof if you decide to do this.

FEATURED ARTIST

Margo Selby

Margo Selby is one of the UK's best-known woven textile artists and designers. She makes wall hangings and installations as well as a range of products based on her designs.

'Margo has established an expansive approach to working in textiles, across art, design and industry. Margo's practice as a weaver is at the heart of all of the studio projects. Throughout her work, Margo explores the relationship between man and machine, hand and industry, craft and technology. Alongside her hand-woven art practice, Margo oversees the work of the Margo Selby Studio, designing for industry partners and the production of a wide range of textile applications.

'Colour is the impetus of Margo's work: the effects of colours in combination; the relationship between colour and woven thread; and the impact of colour in a design. Although Margo's approach is experimental, weaving is an extremely methodical practice. There is a close affinity between Margo's weaving, graphic design and architecture, in that they are all precision disciplines. Her approach to colour was recognised in 2021 with the award of the Turner Medal for Britain's Greatest Colourist.

'Margo has worked continuously on building her practice as a textile artist, developing her hand-woven artworks with increasing scale and ambition. These artworks explore how colour is constructed in woven cloth, and are woven by hand on electronic dobby looms in Margo's Whitstable studio. They often use a lampas weave structure, which allows pure weft colour to sit on pure warp colour. This structure was first developed by French and Belgian Huguenot weavers, transforming the English textile trade of the seventeenth century. Margo uses these structures with a modernist design sensibility and contemporary outlook, creating these artworks as if "painting with yarn".

'When putting together colour for her artworks, Margo often references the ideas of pointillism and divisionism, where colours mix in the eye to create new colour. Ideas are developed through the creation of yarn windings, the process of wrapping yarns around cards and blocks to determine stripe, proportion and contrast along with exploring how warp and weft will interact.

'Margo's hand-woven artworks have been exhibited internationally and have been created for private individuals, galleries, architects, interior designers and cultural institutions. These works range from small-scale commissions to site-specific projects, using colour and textiles to create an uplifting experience within a space.

'An example of Margo's installation work is *moon landing*, a 16-metre hand-woven site-specific length made for the Stamp Stair at Somerset House as part of a collaborative project with composer Helen Caddick. The *moon landing* textile is a celebration of the mathematical and technical possibilities of weaving, and the crossovers of pattern, tone and rhythm found in both music and woven textiles.'

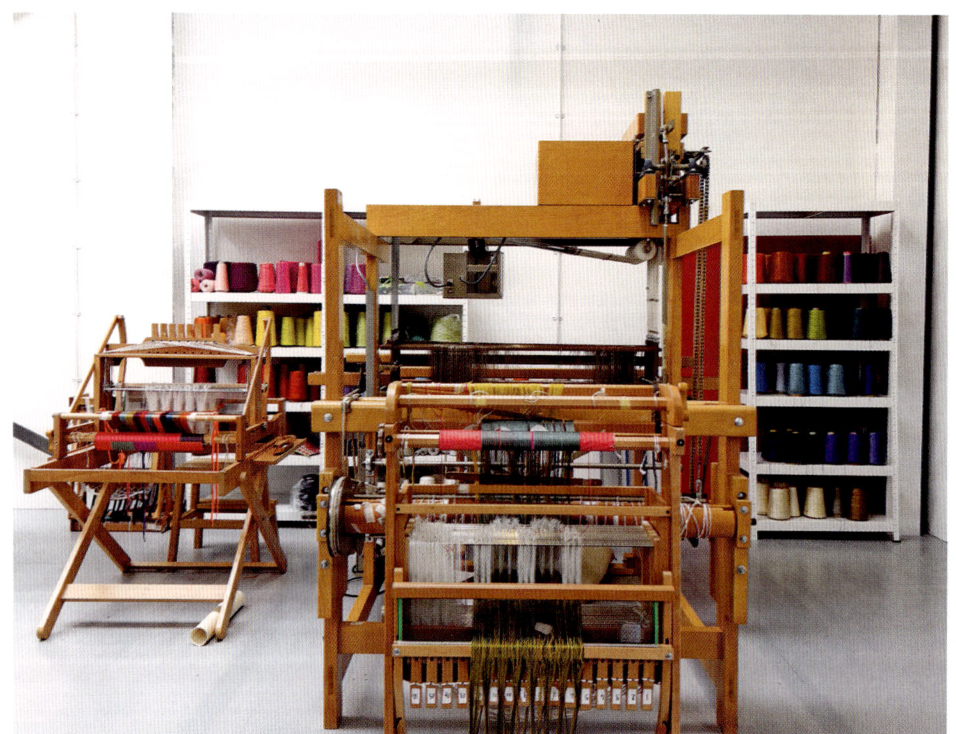

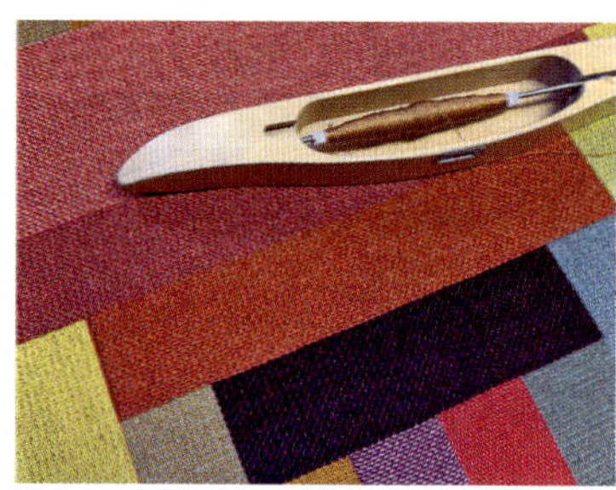

Left: Margo Selby, studio image

Above: Margo Selby, Weaving the *Nexus* series (detail). Woven textile

Opposite: Margo Selby, *moon landing* installation. Woven textile

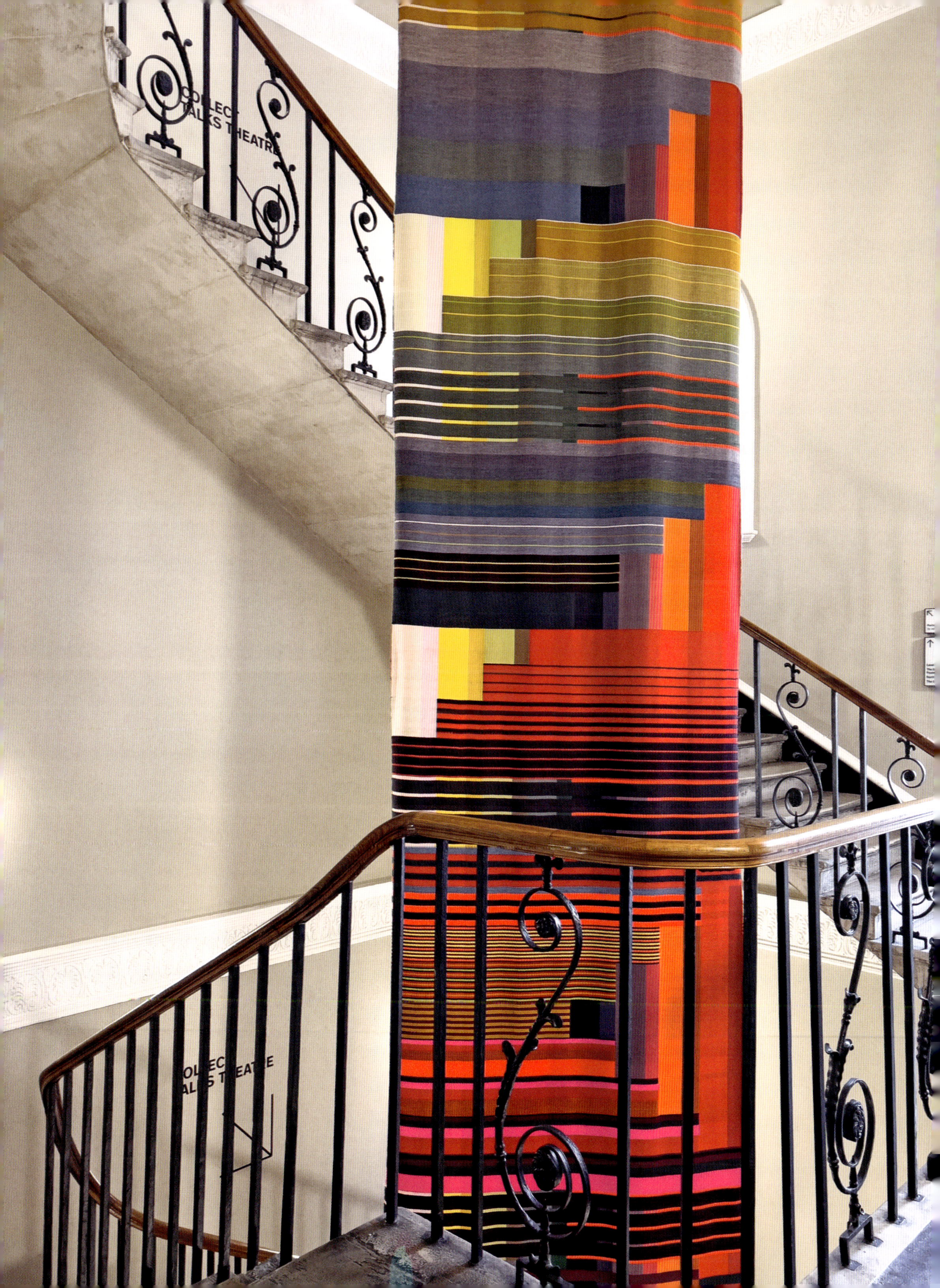

FEATURED ARTIST

Ruth Fox

Ruth Fox is a digital surface print designer and artist. She trained in the UK and is now based in Victoria, BC, Canada. She describes her work here:

'My most recent body of work is inspired and informed by my manipulation of handmade cross-stitch patterns. During the pandemic, I started creating samples that were then photographed and transformed into repeat patterns. The pixelated effect of enlarging handmade designs is kept and adds to the "handmade" look. My colours are strong and harmonious but not always expected. I have made a range of products using these prints in a variety of materials.

'What I find particularly interesting working with digital is that it can actually enhance the handmade craft element of a design and make it more relevant and contemporary to modern-day design. As textile designers, we are constantly taking inspiration from the past: whether it's an aesthetic or just a technique, traditional craft methods hugely inform our practice. What I like about digital print is that it allows this "craft look" to be updated and recycled.'

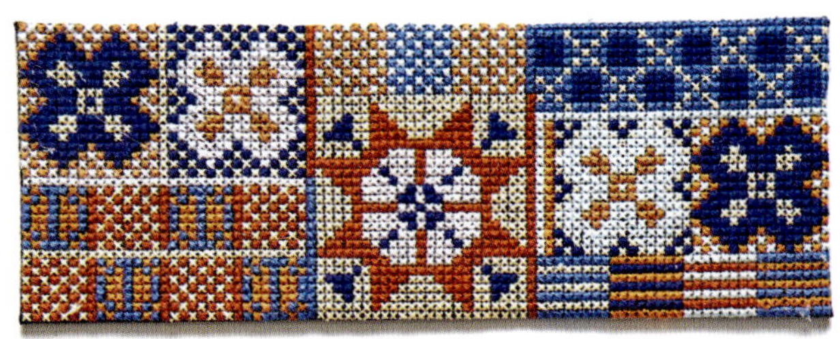

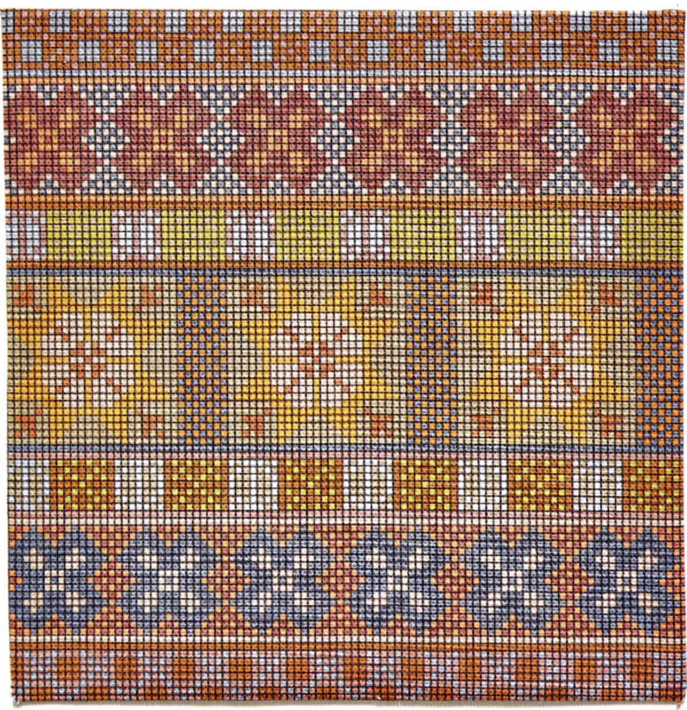

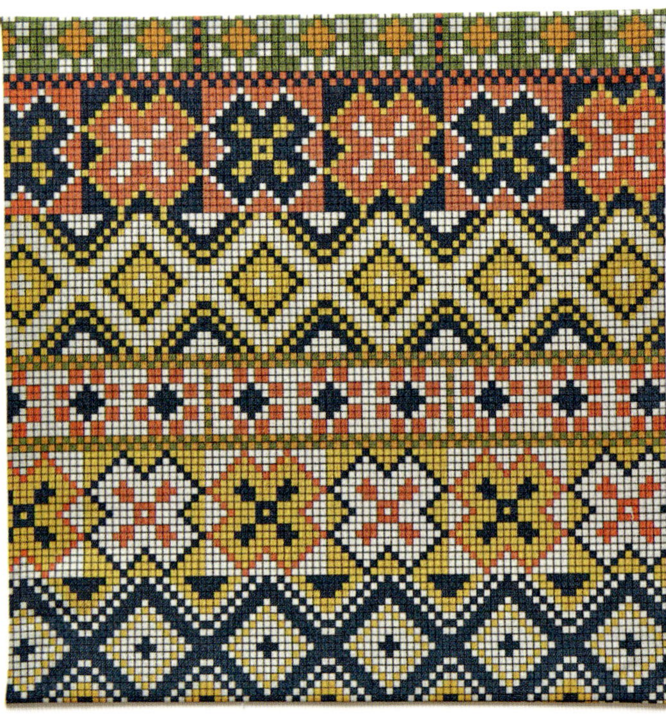

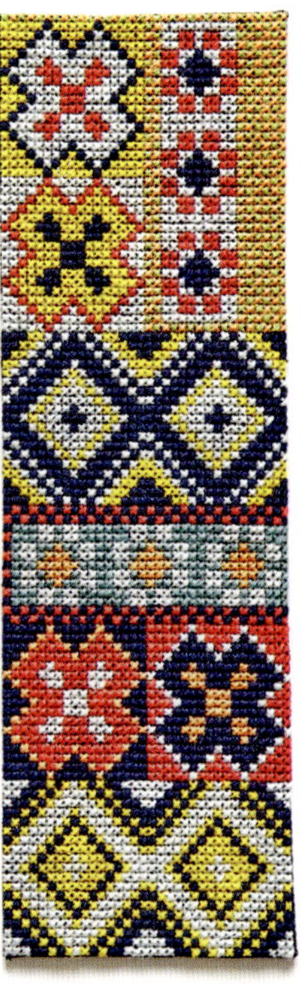

Above: Ruth Fox, *Digital Print on Cotton Fabric*

Left: Ruth Fox, *Digital Print Sample*. Digital print and cross stitch on binca/aida

Digital media has become accessible and useful for the hobby practitioner in recent years. Once the preserve of professional print designers, it was expensive, and required training to create a design. Now, however, many sites include design tools to help you, and it is a great way of creating repeat patterns and abstract designs that can be used to make or cover objects. You can use a relatively simple design to create a complex pattern and have it printed on a range of fabrics quite quickly.

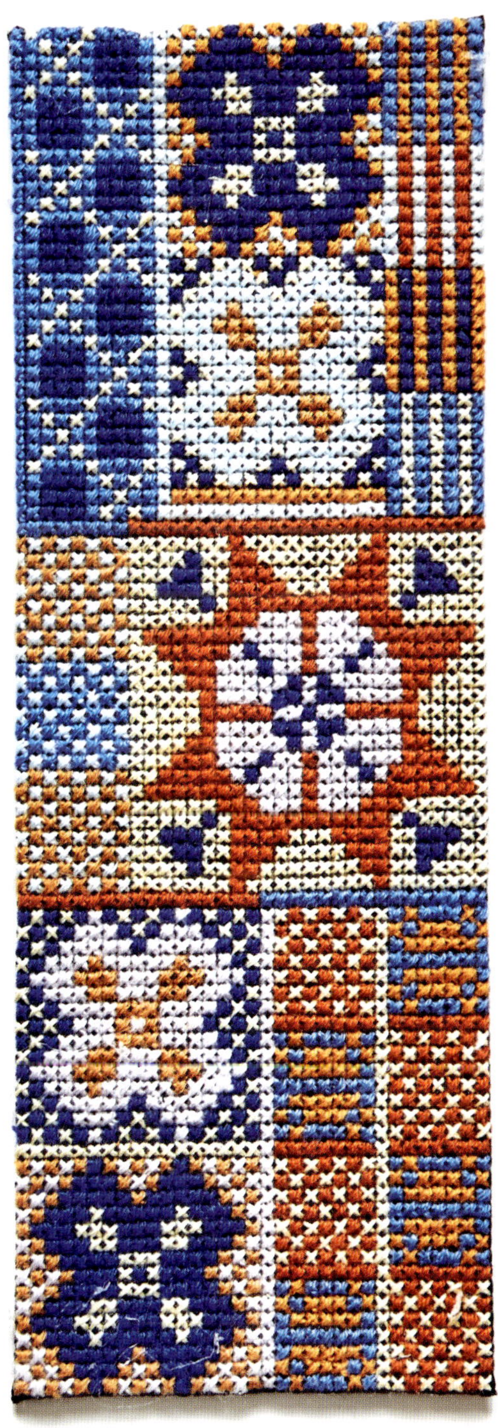

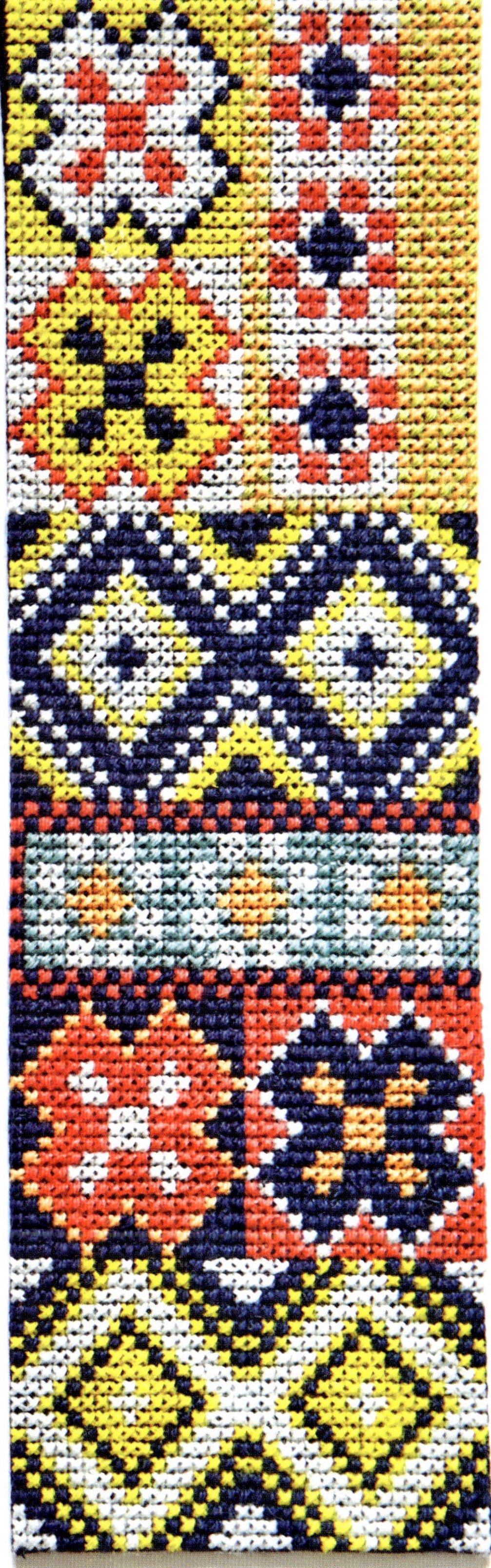

Right: Ruth Fox, *Cross Stitch Samples*. Cross stitch on binca/aida

WORKSHOP

Making a stencil and screen printing

Traditionally, textile printing was also done with screen printing and stencils. I like to use screen prints for background patterns and designs. It is a great way to create texture and pattern in a relatively quick way.

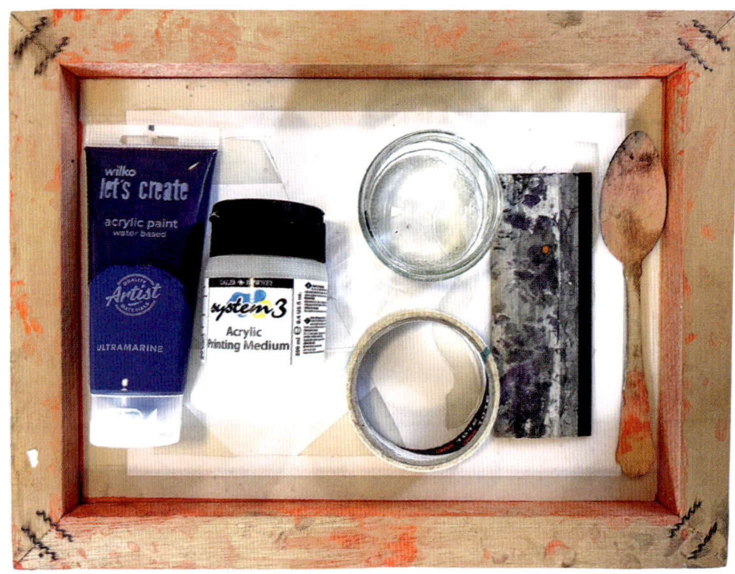

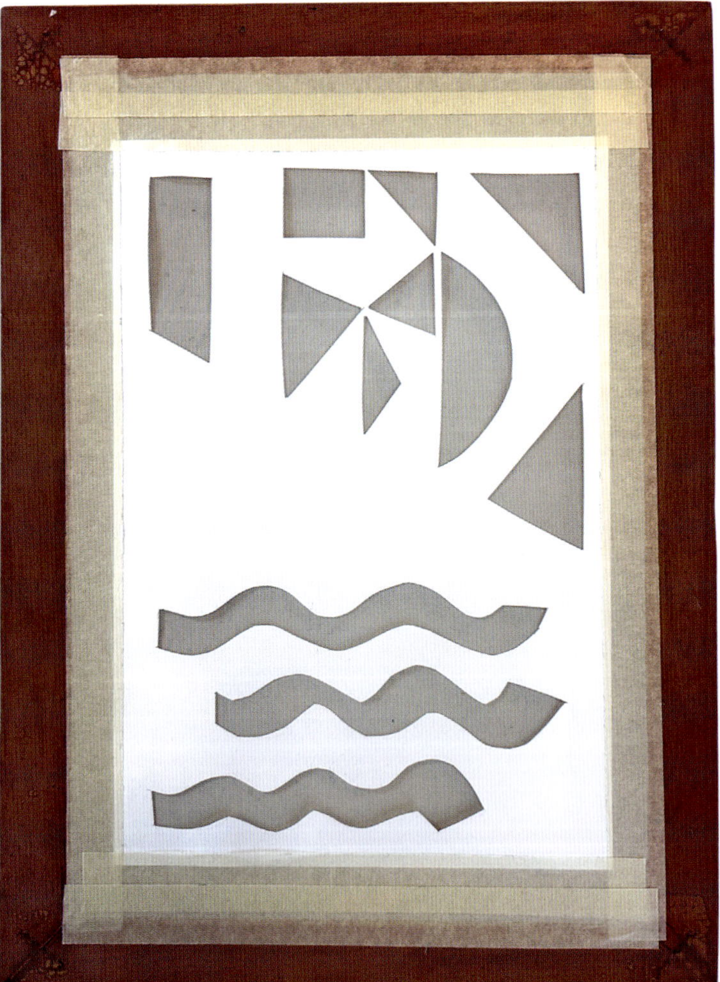

MATERIALS REQUIRED

- Screen (small to start; mine is A4 size; 8½ x 12in)
- Squeegee
- Screen-printing ink or fabric paint made with fabric medium and acrylic paint
- Thin cartridge paper, masking tape
- Scalpel and cutting board
- Fabric to print on

METHOD

1. Draw your design on the cartridge paper and cut out, leaving space for the ink to go through with each cut. Tape it to the screen on the outside, covering any blank spaces.
2. Tape your fabric to a flat smooth surface. Cover it with newspaper or plastic to protect it from the ink.
3. Place your screen over the fabric and spread a thin line of ink across one edge of the inside of the frame and mesh. Spread with a brush or stick if necessary.
4. Using the squeegee, gently pull the ink across the stencilled image. Repeat on another space if you want to continue the image. Protect the wet print with a piece of paper.

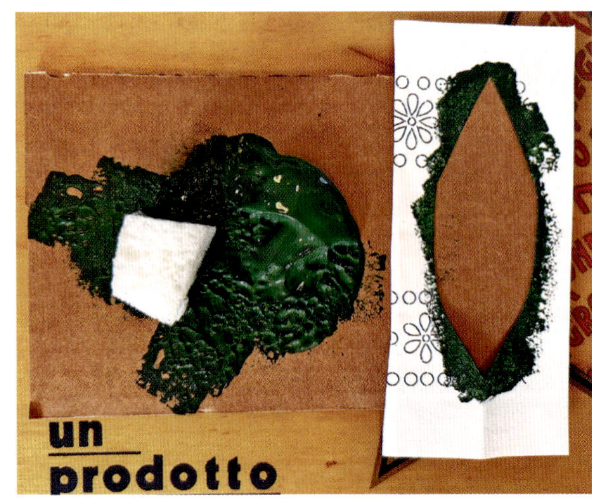

Opposite top: Screen-printing essentials

Opposite bottom: Screen and stencil

Right: Screen print drying

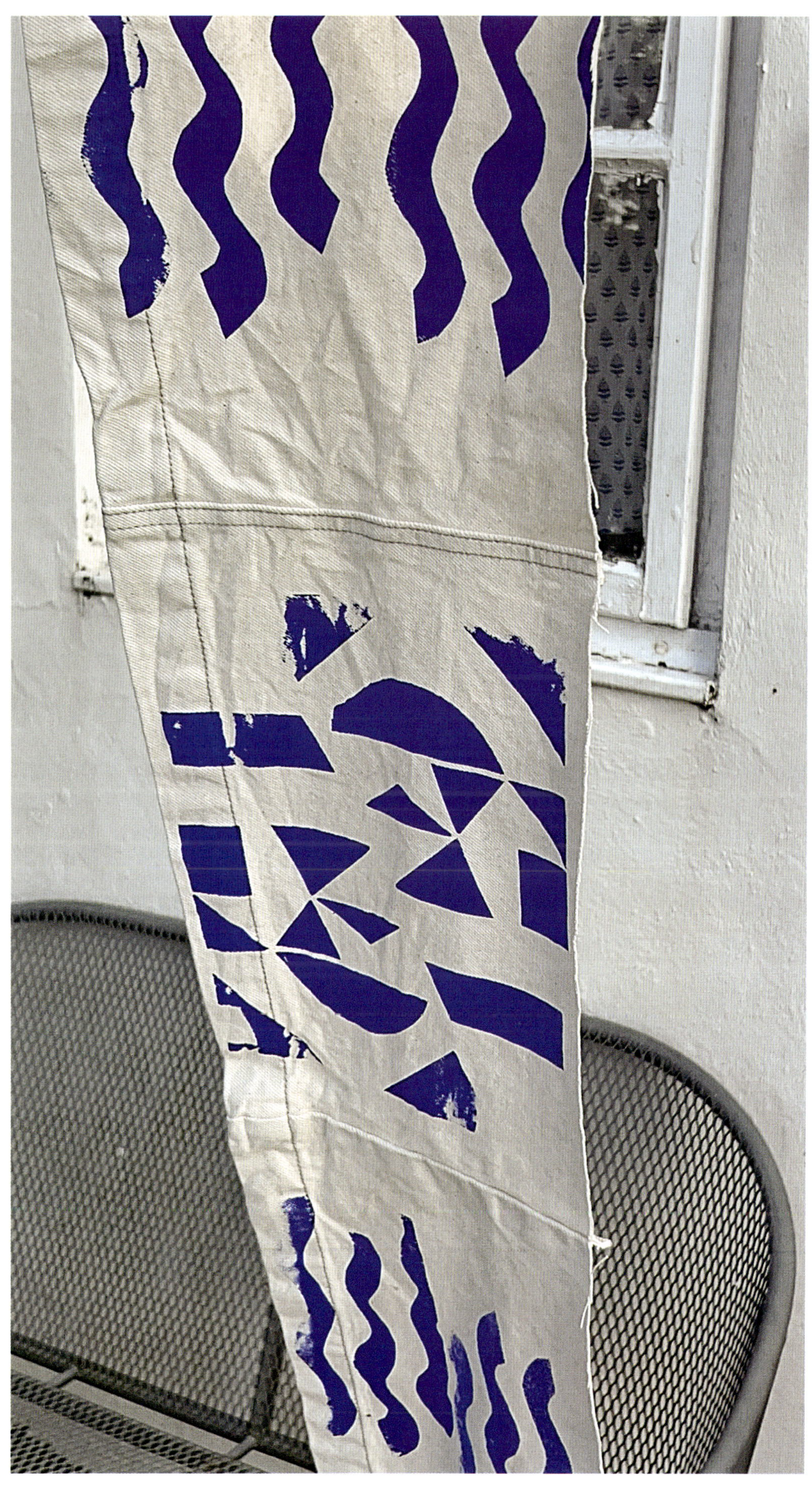

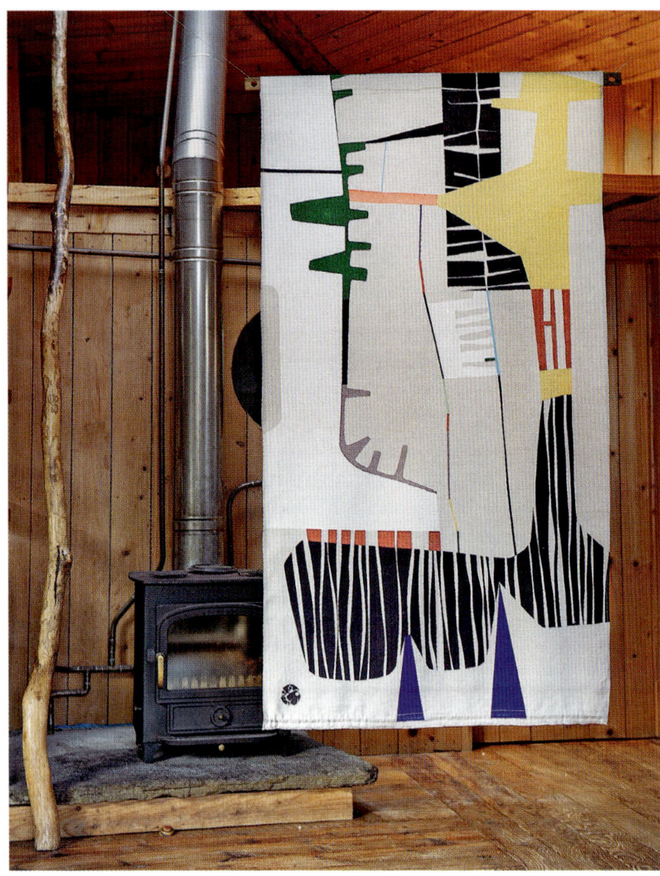

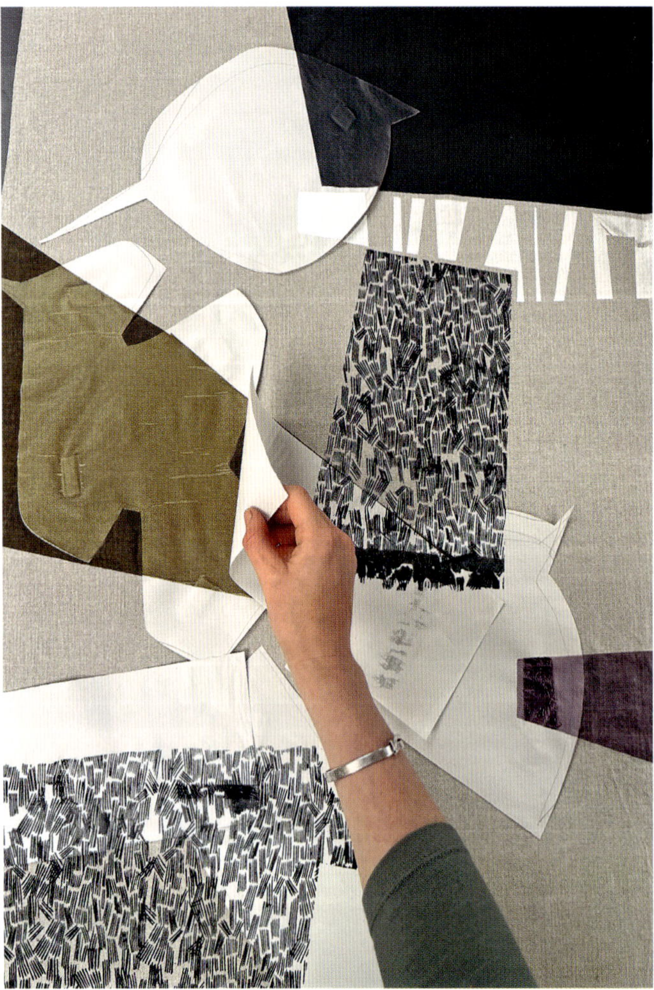

FEATURED ARTIST

Eve Campbell

With an interest in creating surface pattern inspired by Scottish nature and architecture, Eve Campbell creates printed textiles, wall hangings and ceramics. Through paper stencilling and screen printing, her prints capture the colours, shapes and patterns of nature on Scotland's West Coast. She describes her work here:

'After graduating in Textile Design from the Glasgow School of Art in 2018, I established my print studio in Tighnabruaich, Argyll. From there, I create my textiles, interpreting nature in abstract form for homes and spaces. My signature pieces are my large-scale screen-printed wall hangings.

'Using a natural unbleached linen base, I apply bold colours layer by layer using paper stencilling and screen printing. My printing process is unique due to this paper stencilling technique, where each shape and pattern is cut from newsprint before being printed. This process allows me to design as I am working on the print table, responding and reacting to each layer as I go. The fleeting existence of each stencil results in each wall hanging being unique and a one-off.

'My inspiration comes from my surroundings: the remote, wild, natural landscapes on the West Coast of Scotland and the interaction between nature and humankind. My creative process begins with a drawing series, often in sketchbooks in the natural environment itself, whether it is the woodland, the shore or the sea. These drawings and sketchbooks sit open on the table and serve as a vital reference point as I print, responding to the shapes, patterns and colours to build up the layers of my wall hangings.'

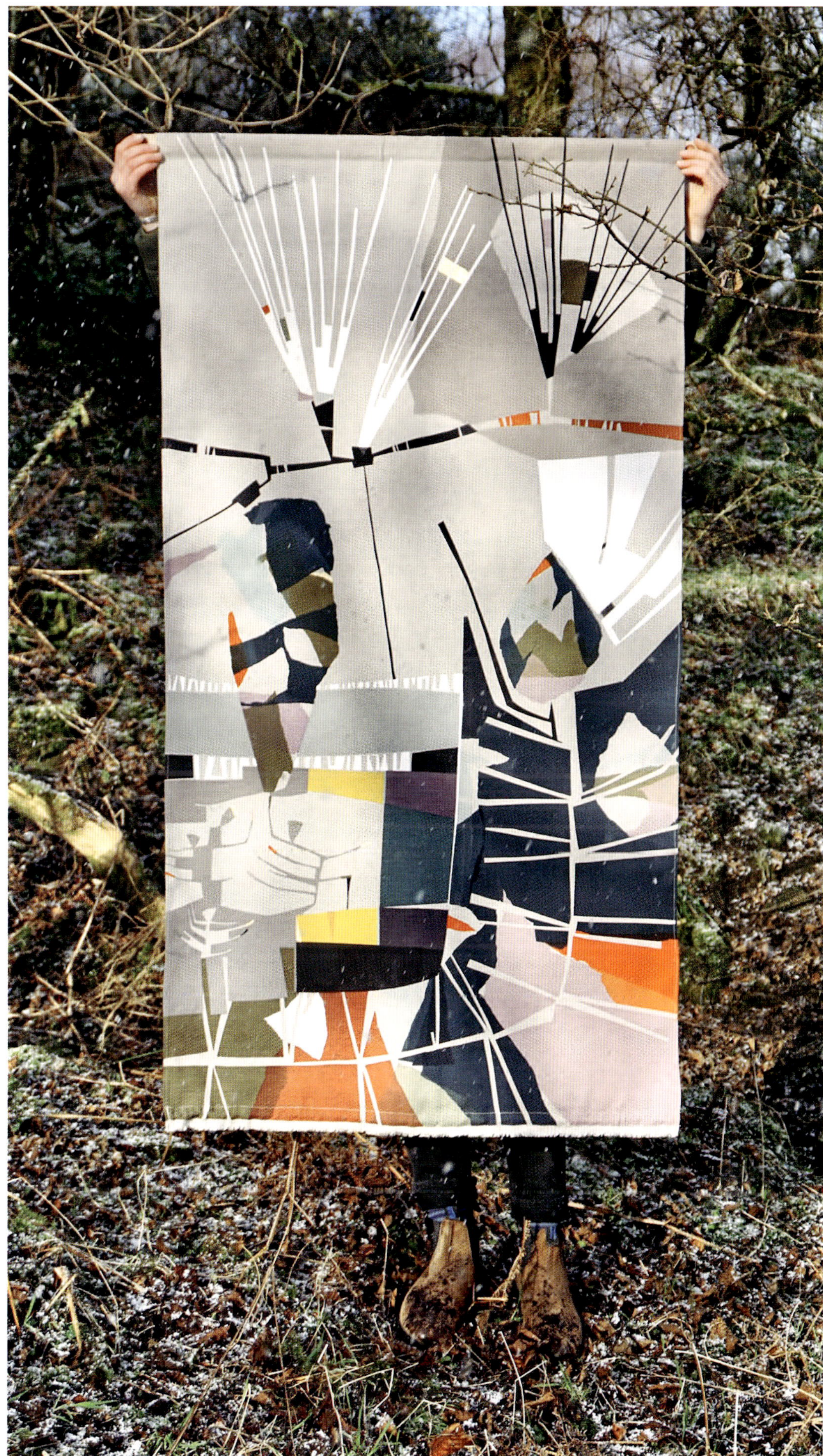

Opposite top: Eve Campbell, *Wildwood*. Printed textile

Opposite bottom: Eve Campbell, *Work in progress*. Printed textile

Right: Eve Campbell, *Evergreen*. Printed textile

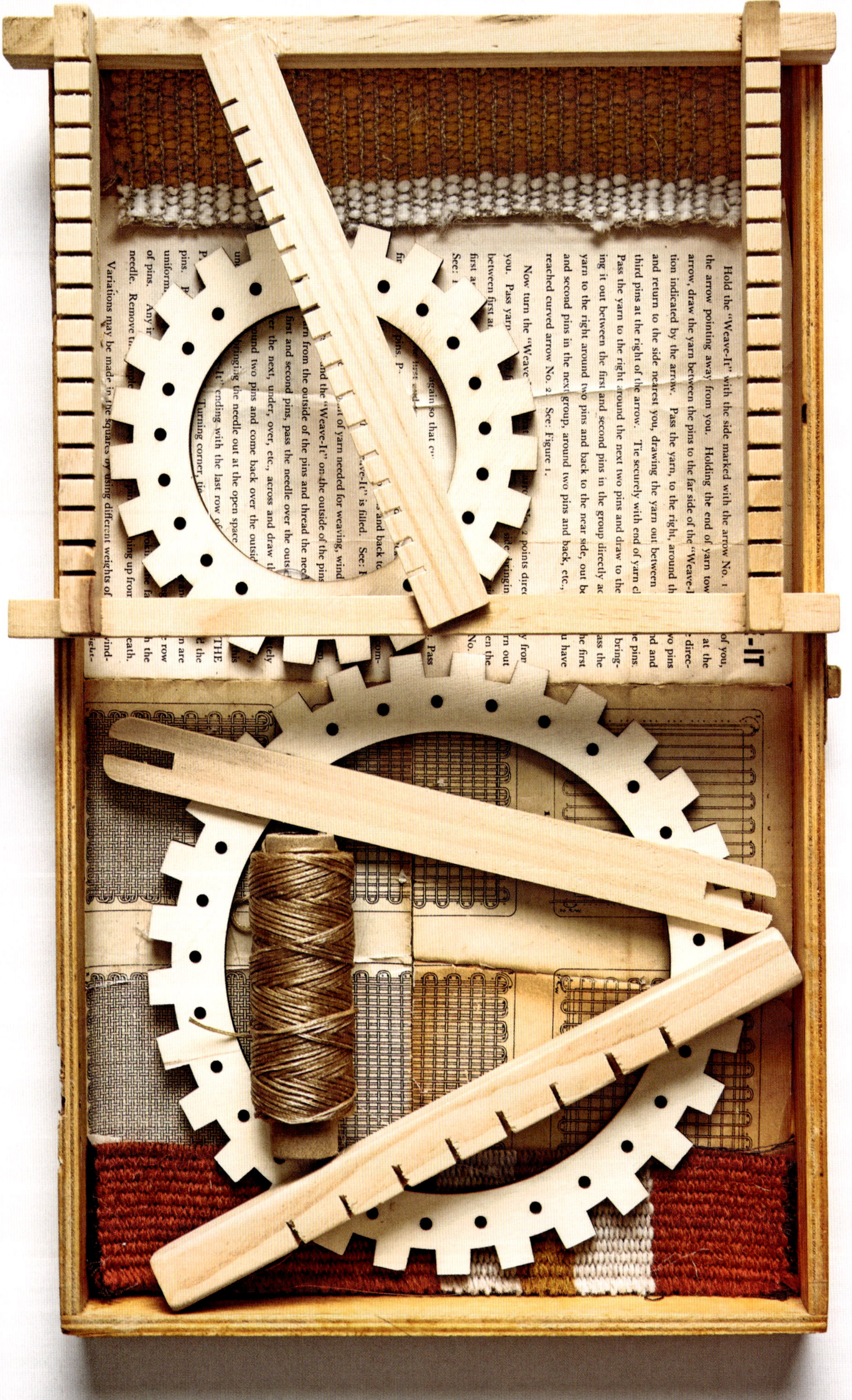

Chapter 4

FOLLOWING THE LINE

'Abstraction is real, probably more real than nature.'

Josef Albers (1888–1976)

Opposite: Anne Kelly, *Inspiration Box, Weaving*. Mixed media in wooden box

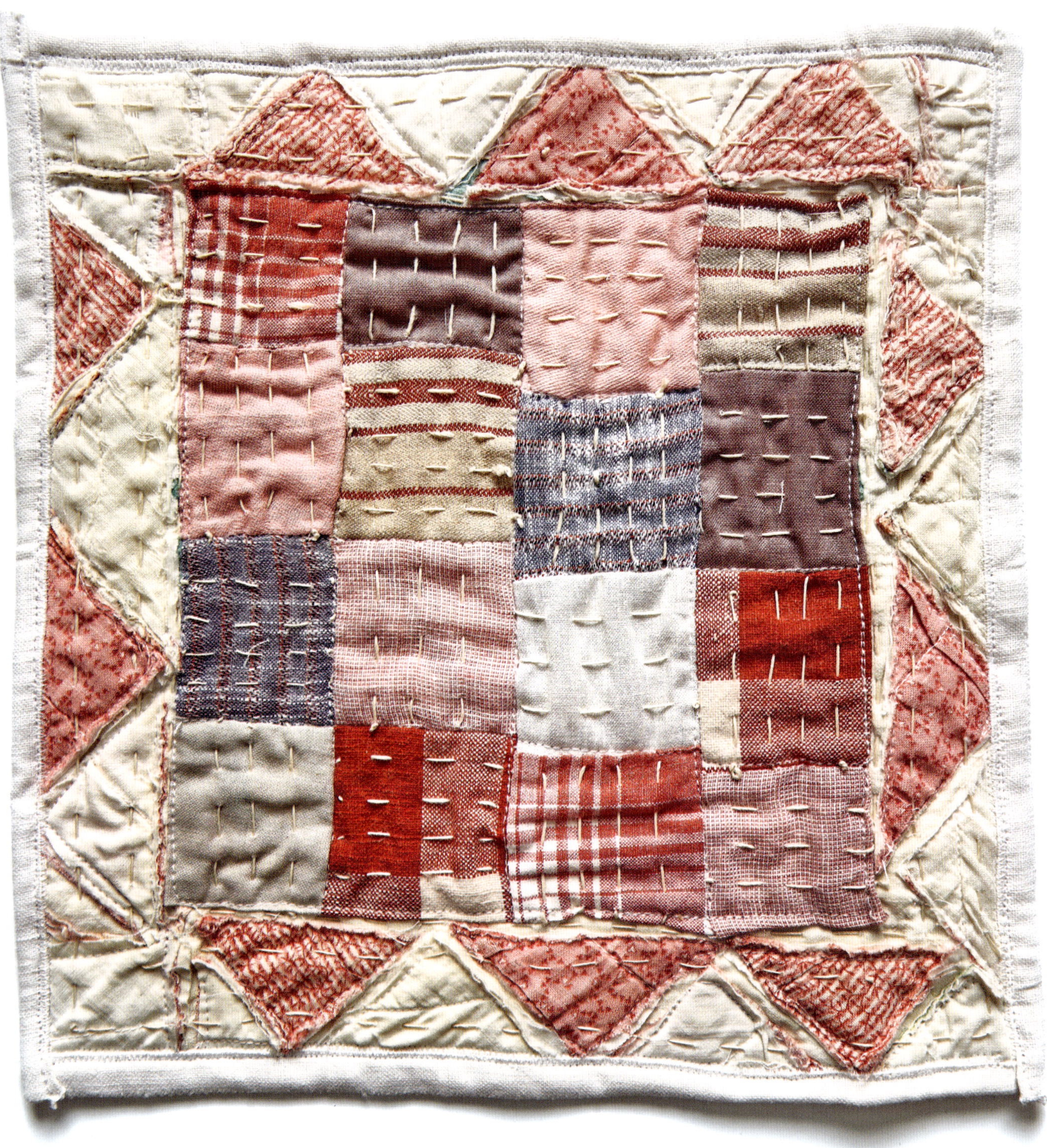

In Chapter Four, we look at lines and linear compositions. Line is a powerful tool to use in all forms of art, but you could argue it is more necessary in textile art. It provides a visual and essential constructive element, joining and layering fabric together. The artists selected for this chapter all use lines in different forms.

We also explore the traditional stitching techniques of Kantha and Boro. These have become present, and some could argue over-used, in recent years with the popularity of 'slow stitching'. They are easy and useful for everyone to use, however, and work well with our projects in this part of the book.

PATCHWORK SAMPLES

I first began quilting when I was a young, bored teenager, and had soon corralled my mother's Singer sewing machine. I have often used quilting techniques in my practice, for backgrounds and sourcing vintage quilt tops for textile collage. I wanted to experiment with abstract shapes and try to create some small samples.

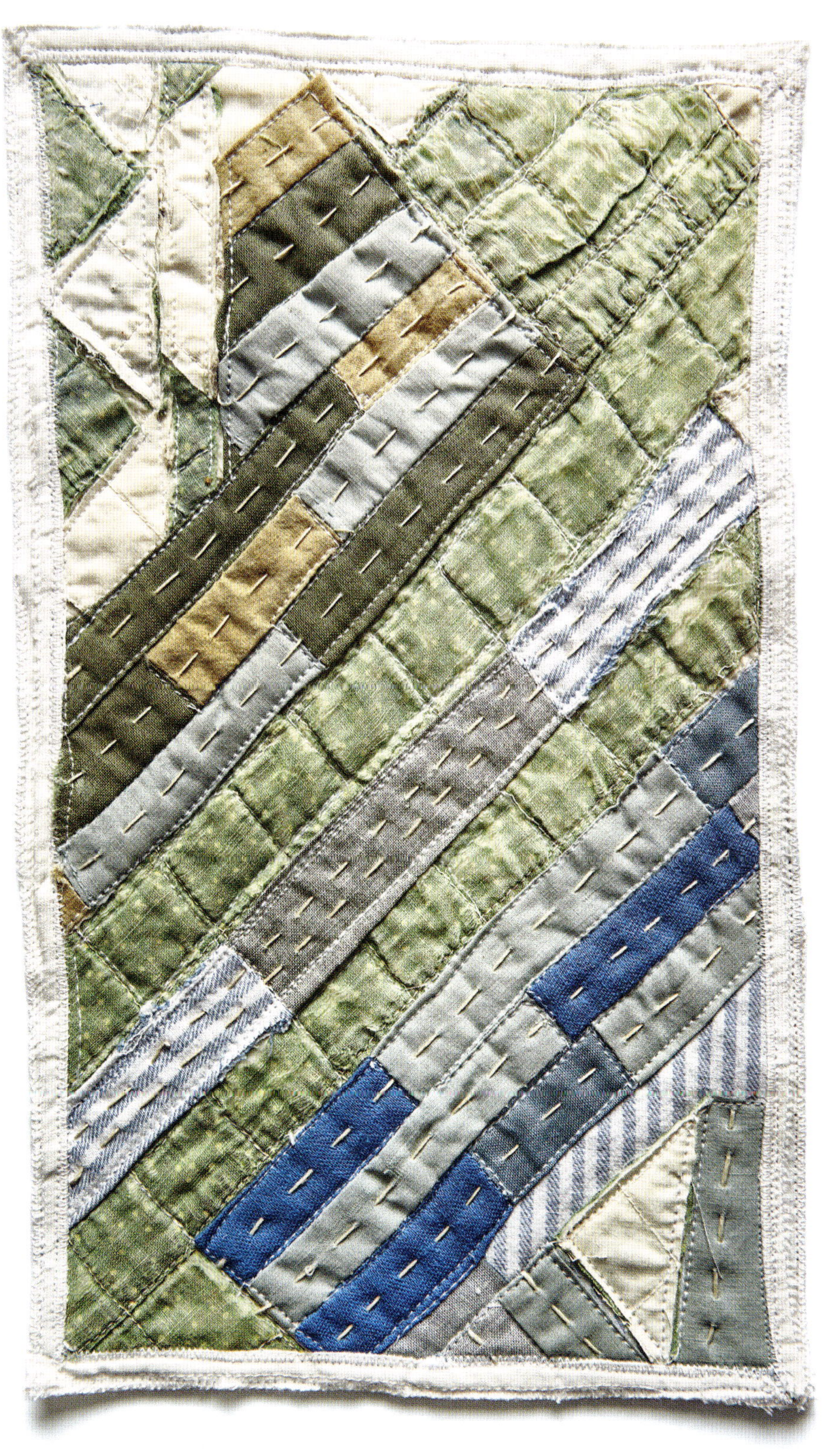

Opposite: Anne Kelly, *Small Pink Quilt*. Quilted stitched textile

Left: Anne Kelly, *Small Green Quilt*. Quilted stitched textile

WORKSHOP
Making a small patchwork sample

MATERIALS REQUIRED

- Cotton background/base fabric no larger than A4 size (8½ x 12in), any shape
- Strips of plain fabric for patchwork, no wider than 5cm (2in) in range of colours
- Wadding (batting) for centre; you can also use blanket cloth
- Thread for machine (if desired) for quilting, embellishing
- Iron and ironing board

METHOD

1. Choose fabrics and colours that work well together. Cut and arrange the pieces in the order and pattern in which you would like to stitch them. You should have enough to cover the base with a 1cm (½in) seam allowance.
2. Stitch the pieces together in the traditional face-to-face manner, leaving a seam allowance as above. When you have completed an area or strip, iron the seams flat and open.
3. When you are ready to join the top to the base, tack (baste) stitch the top to the wadding (batting).
4. You can now quilt your piece, by hand or machine. Create new patterns by adding lines inside and outside the seam lines. Contrasting colours of stitching can also be effective. I used a heavyweight linen thread to hand quilt with.
5. Flip the piece over and stitch it to the backing fabric (face to face), leaving a 1cm (½in) border in a U-shape (leaving one side open).
6. Trim the seams so that they are half the original size; you can trim diagonally on the corners. Then fold over the top and the base fabric slightly and join together with stitch.

Below left: Anne Kelly, *Colour Study*. Watercolour on paper

Below right: Lesley Patterson-Marx, *Quilted Paper Card*. Mixed media on paper

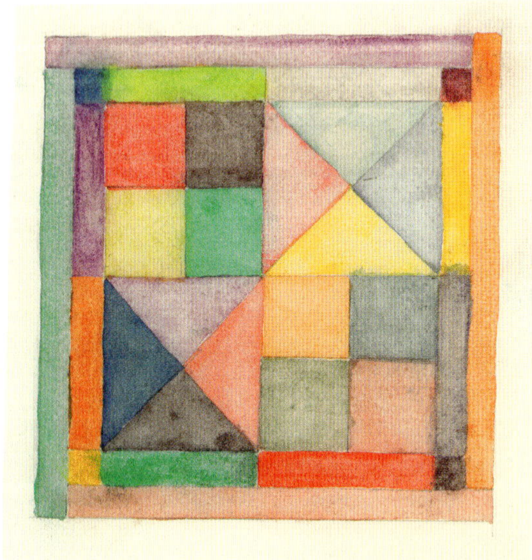

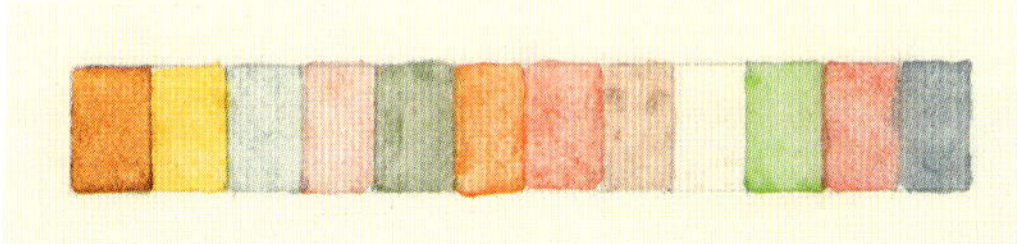

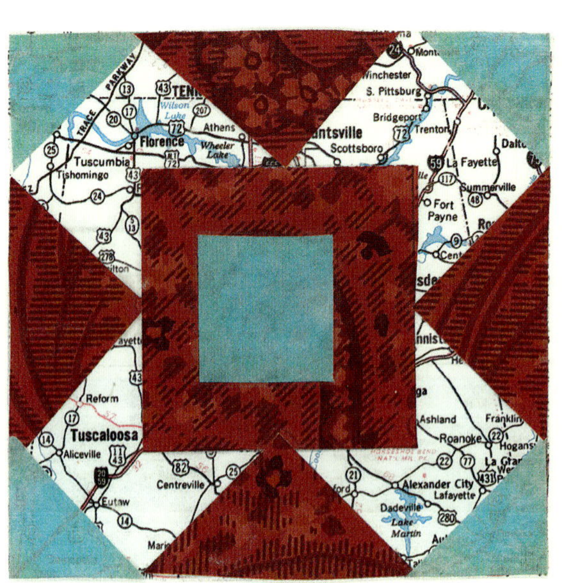

AN ABSTRACT PATCHWORK SAMPLER

I like to use recycled and donated fabric in my practice. A student gave me a collection of woven wool samples; I wanted to make something with them and decided to create a simple patchwork sampler. I joined the pieces together diagonally and then in strips. I used an old piece of patchwork quilt as a base and backing fabric.

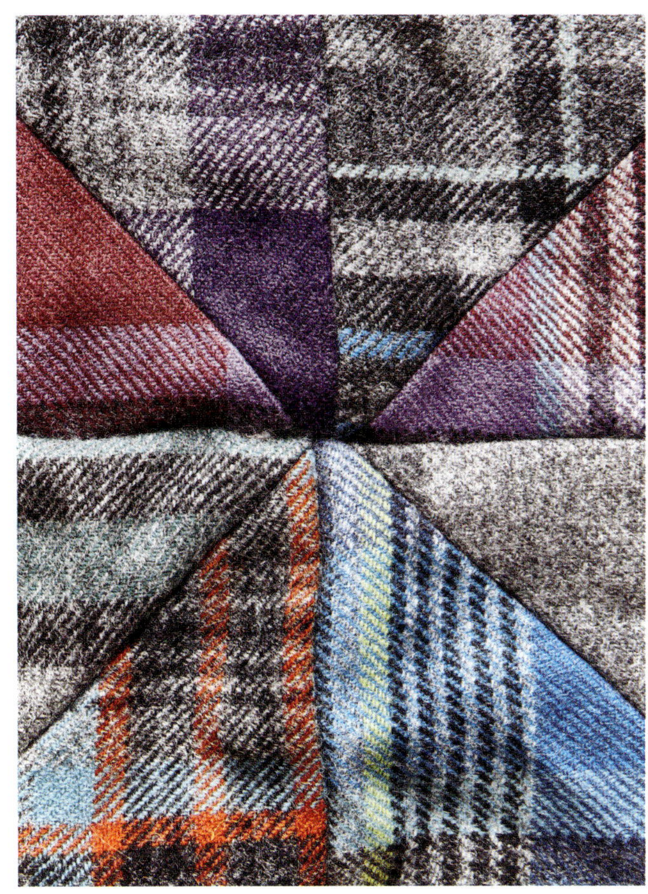

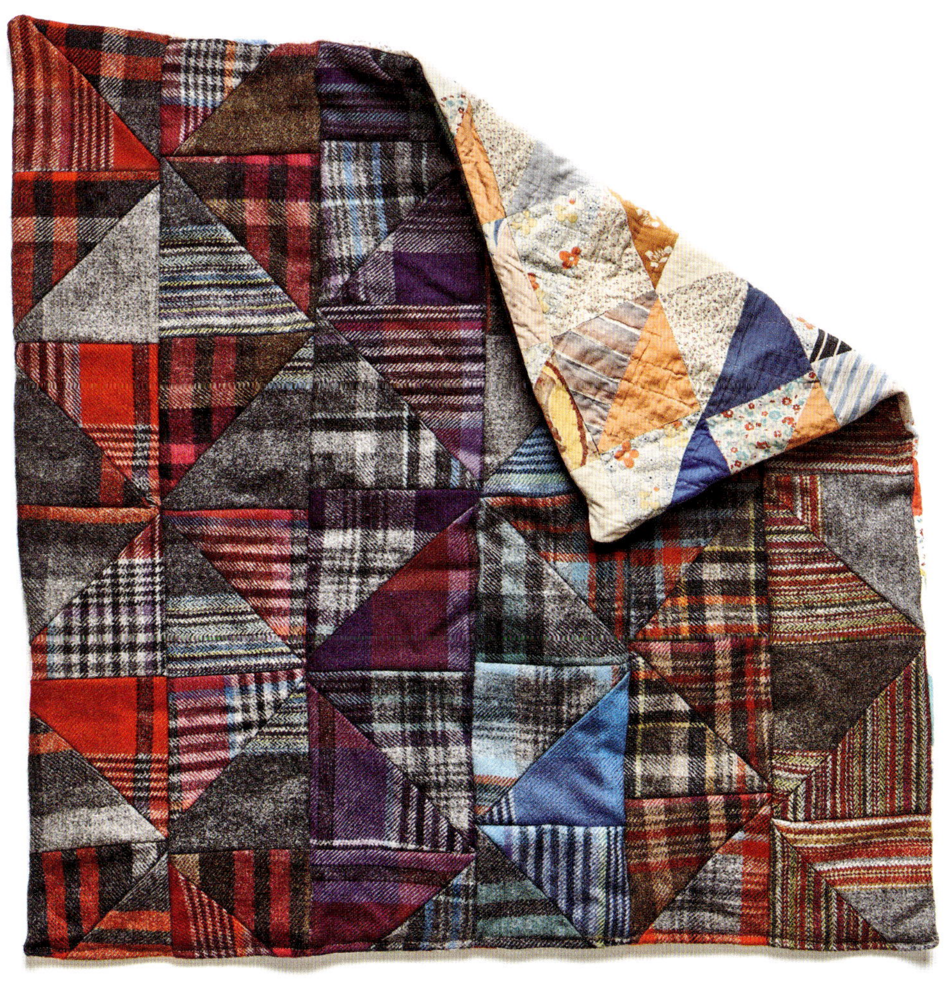

Above: Anne Kelly, *Wool Patchwork Quilt* (detail). Mixed media quilt

Left: Anne Kelly, *Wool Patchwork Quilt*. Mixed media quilt

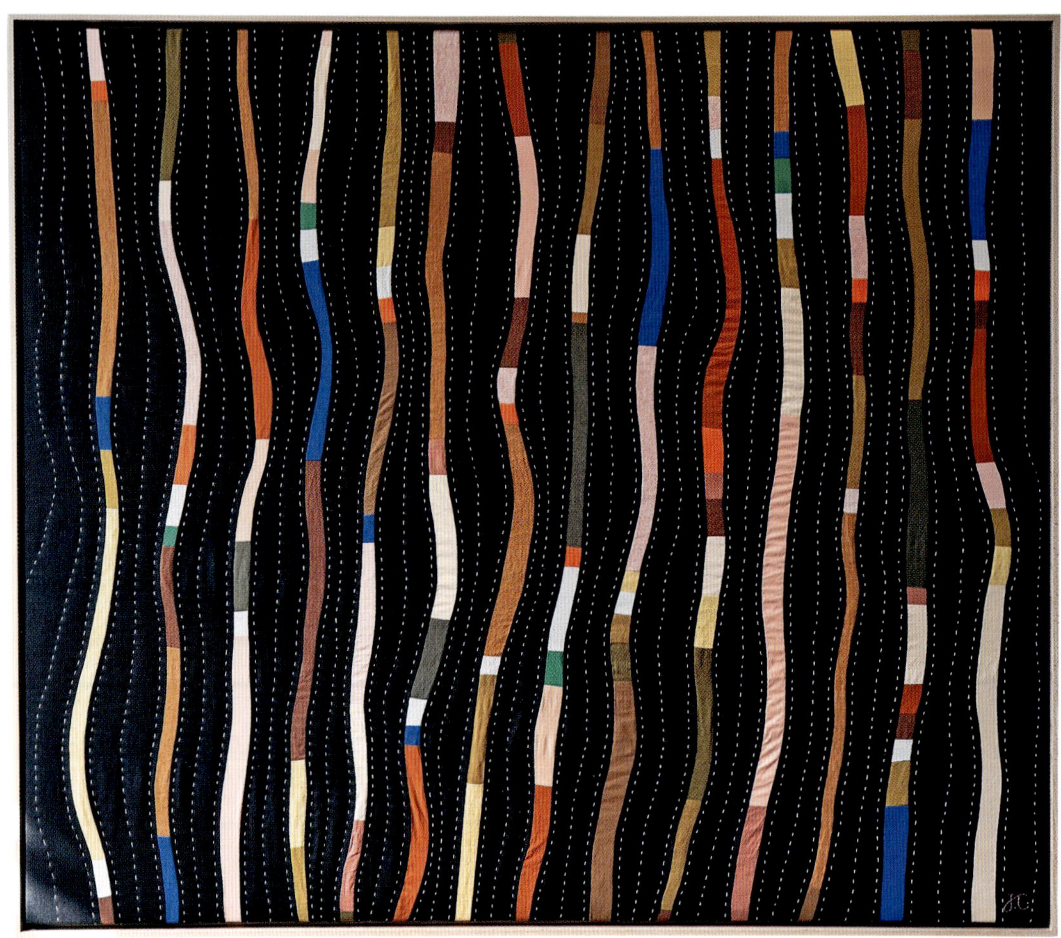

Left: Jessie Cutts, *Night Lights*. Patched and stitched framed wall hanging

Below: Jessie Cutts, *Colour Blocks Heavy Stitching*. Patched and stitched framed wall hanging

Opposite: Jessie Cutts, *Trimmings and Offcuts* series. Patched and stitched unframed pieces

FEATURED ARTIST

Jessie Cutts

Jessie Cutts makes distinctive geometric and linear quilts. They are striking in colour and by design, as she tells us here:

'I am a textile artist using age-old, traditional patchwork and quilting to make abstract textile works. I favour organic forms and free-form cutting and sewing, taking an improvisational approach to pattern and line work. These are "made" artworks, using a domestic, functional craft to create pieces that are at once modern, colourful and textural. I work in direct response to the fabric, by mostly free-form sewing – using trimmings from previous works, scraps, upcycled clothes and offcuts from clothing manufacture.

'I stumbled upon this way of working with freedom and not too much planning out of being intimidated by formal quilt patterns. I gave myself room to make "mistakes" until I landed on this approach. For me, using mistakes and improvisation as a positive force keeps my work feeling fresh and unconstrained. I try to make use of even the tiniest pieces of fabric, often taking the already sewn together offcuts as the start of the next piece (this is where my smaller works usually start), and this gives me a chance to work in a very unplanned way. I find myself working on these pieces as a break when I am working on larger works that take longer to complete.'

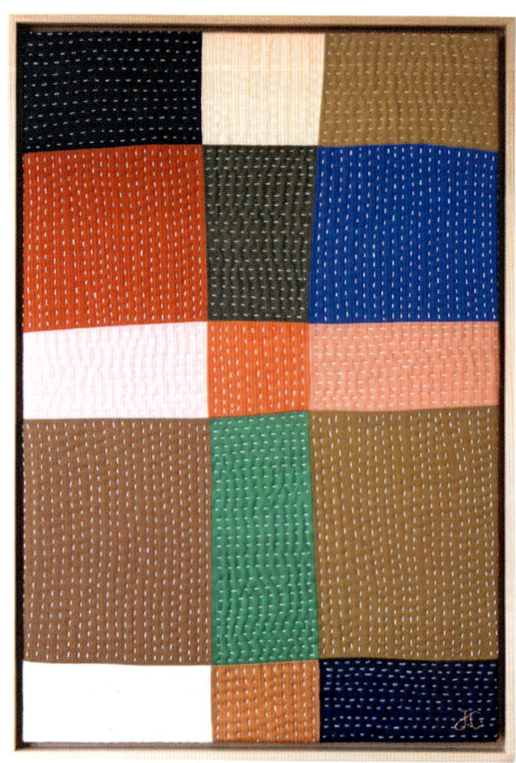

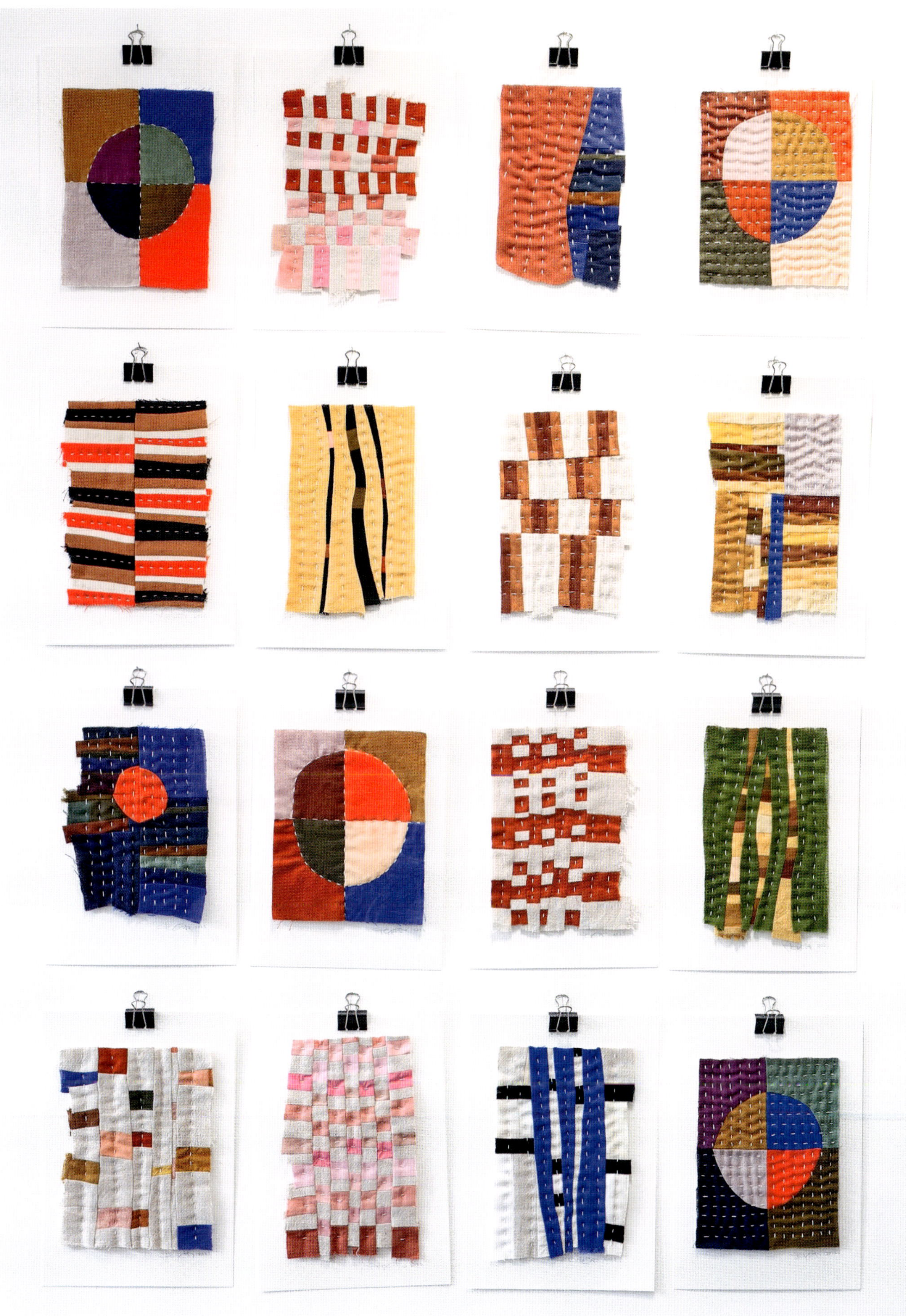

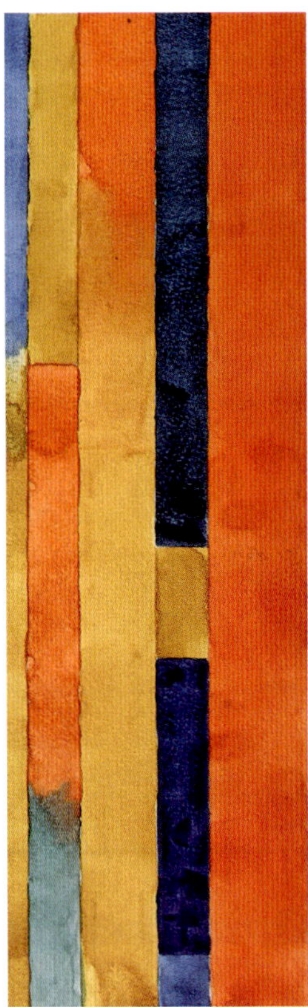

Left: Dalia James, *Yellow, Red, Blue.* Watercolour study

Below: Dalia James, *Yellow, Red, Blue I, II.* Woven textile

FEATURED ARTIST

Dalia James

Dalia James is a weaver who makes rich and complex work, using a large range of materials and influences, as she explains here:

'The relationship between colour and geometry intrigues me, and my work often references early twentieth-century design and artistic movements, specifically those formed during the interwar years: the Bauhaus, De Stijl, Dadaism and Constructivism. Exploring the relationship between the ordered rectilinear blocks and the random nature of the dip-dyed sections of colour have become key themes of my work.

'The paintings of Josef Albers have been a great influence. More recently, his colour theories have taken on greater significance as I think more about how my work is perceived by those viewing it, and how his theories can influence my work. The ethics and integrity of William Morris, founder of the Arts & Crafts Movement, have also been of great importance in terms of my commitment to the craft of hand weaving and my consciousness of the effect of my practice on the environment. My transition from synthetic to natural dyes and my desire to use only natural fibres is in part the influence of Morris.

'The remaining stages of my creative process are done on paper. I use watercolours to create and refine a colour palette using the reference material I collect and then create watercolour studies that help inform how I dye the warp. The draft is also done on paper: I use grid paper to work out the width, length and EPI (ends per inch) of each piece, as well as the threading plan.

'I use a double cloth woven structure for my artwork as this allows me to create vertical and horizontal blocks. These blocks can be wound and dyed separately before the threading process and allow me to change the colours throughout the warp. I primarily use spun silk and bamboo yarns, but I have also used seaweed-based and pineapple yarns. I hand dye all of my yarns, as colour is one of the two key themes of my work. I have never used pre-dyed yarns as doing so would be a compromise. Dyeing is a skill in and of itself. Many weavers do not dye their own yarns, but it is an essential part of my process and key to the integrity of my work.

'Once I have wound the yarns out, I lay each section out on the floor, one next to the other, and use lengths of the weft yarns (which I dye beforehand) to mark out where each colour begins and ends for each of them. These sections are then dip-dyed, primarily to test different colourways of the same design on a single warp. I found that I was most interested in the areas where the colours met. I may mark out the sections, but I never know exactly where one colour will end and where the next will begin due to the nature of dip-dyeing.

'Compositionally, the piece only comes together on the loom. I weave instinctively and do not plan anything other than the warp colours in advance. As I use a double cloth technique, I can choose which warp and therefore which colours to use as I'm weaving. The outcome of the piece is revealed to me only when I take the warp off the loom.'

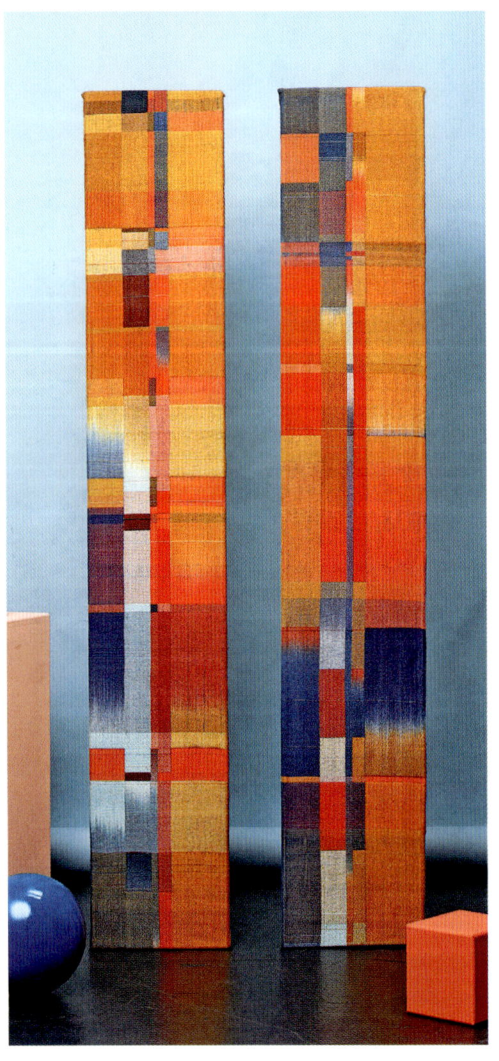

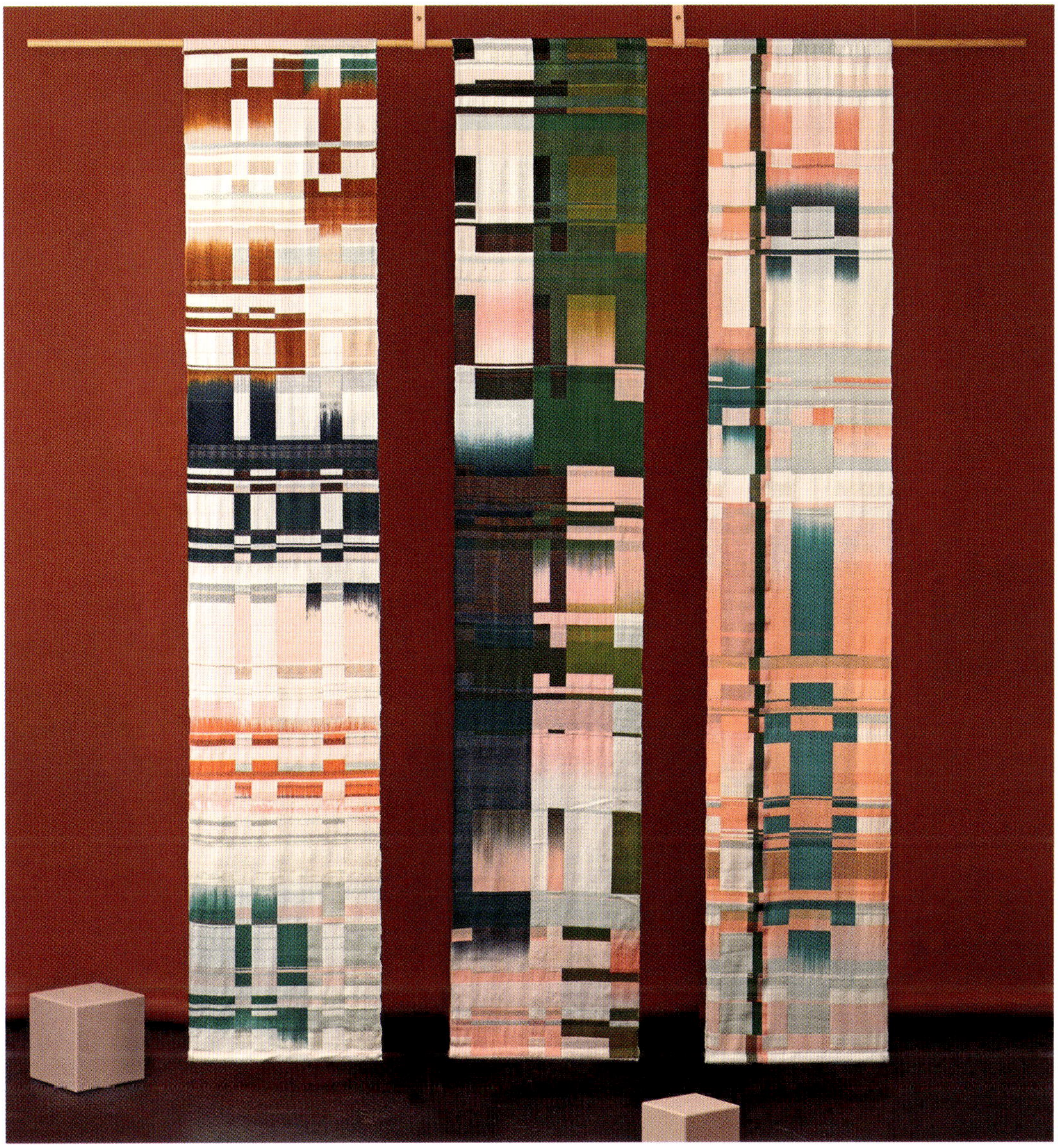

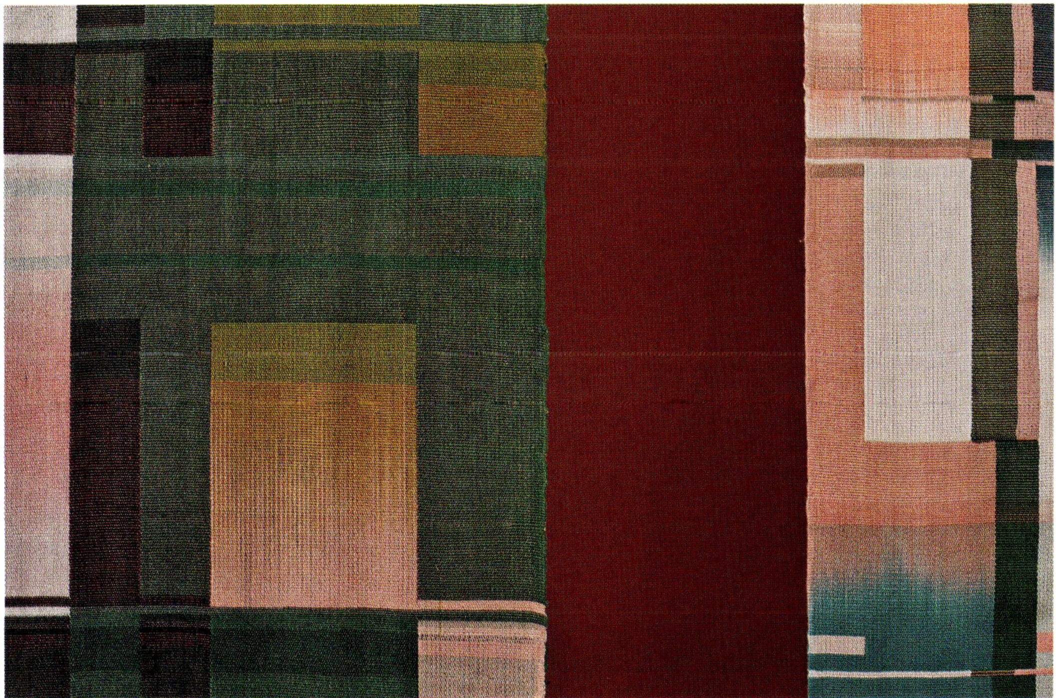

Above left: Dalia James, *Intersection I, II, III*. Woven textile

Left: Dalia James, *Intersection I, II, III* (detail). Woven textile

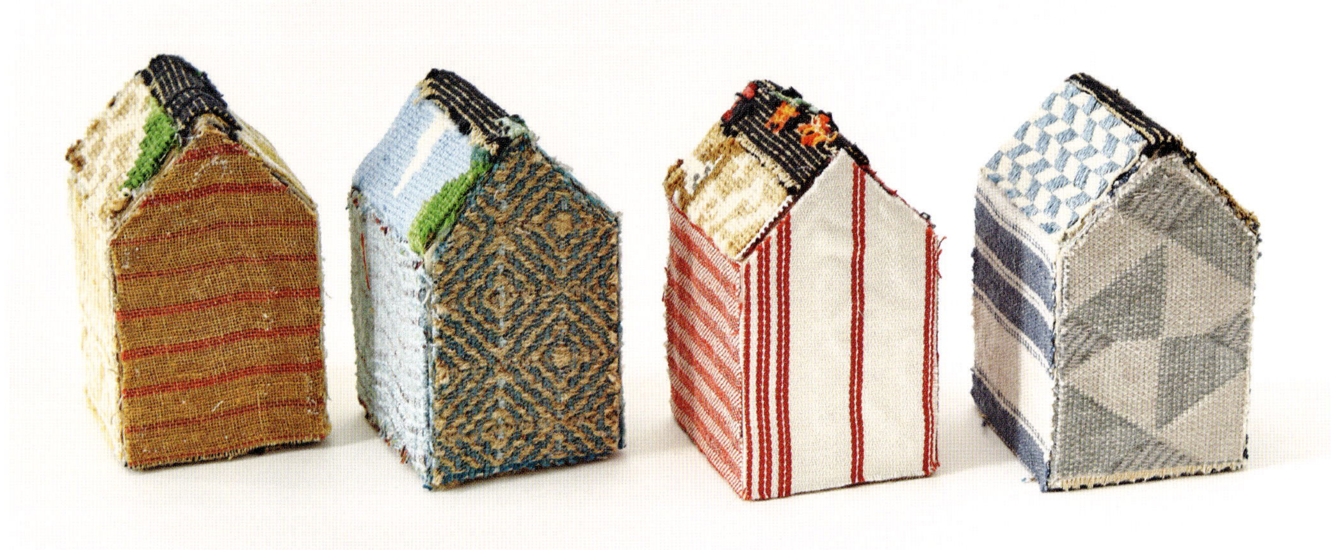

WEAVING

Woven houses

I wanted to showcase different samples of woven fabric, more pronounced than the ones I chose in the sampler, and returned to the house format, using larger wooden house forms and attaching the fabric samples to their sides.

I have also discovered a small vintage loom kit in a charity shop and am using it to create a woven sample. You can make a loom at home with very simple materials, even card: instructions can be found on the internet.

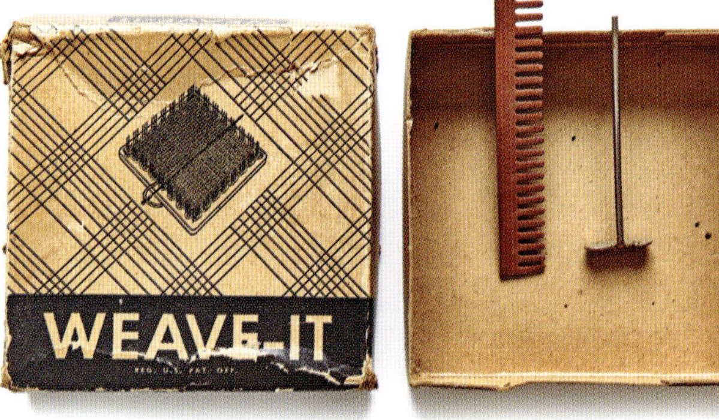

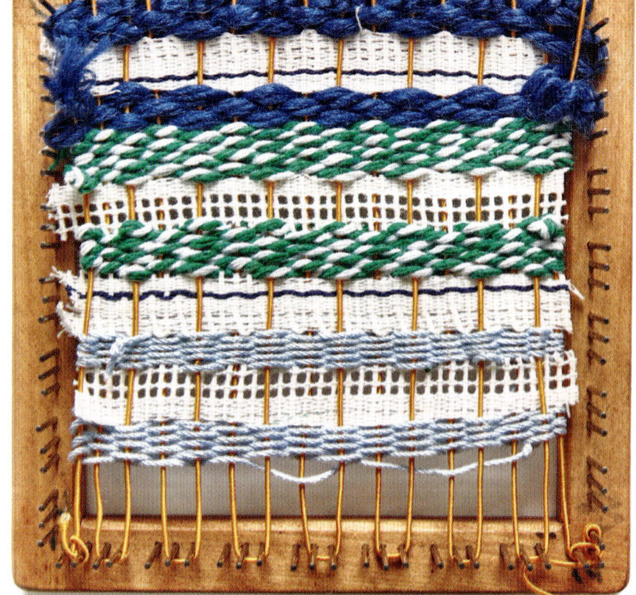

Top: Anne Kelly, *Woven Houses*. Wood base covered in woven textile

Centre: Small loom weaving kit. Cardboard box and tools

Bottom: Anne Kelly, *Weaving Sample*. Mixed media weaving

Opposite: Anne Kelly, *Weaving Sample* (detail). Mixed media weaving

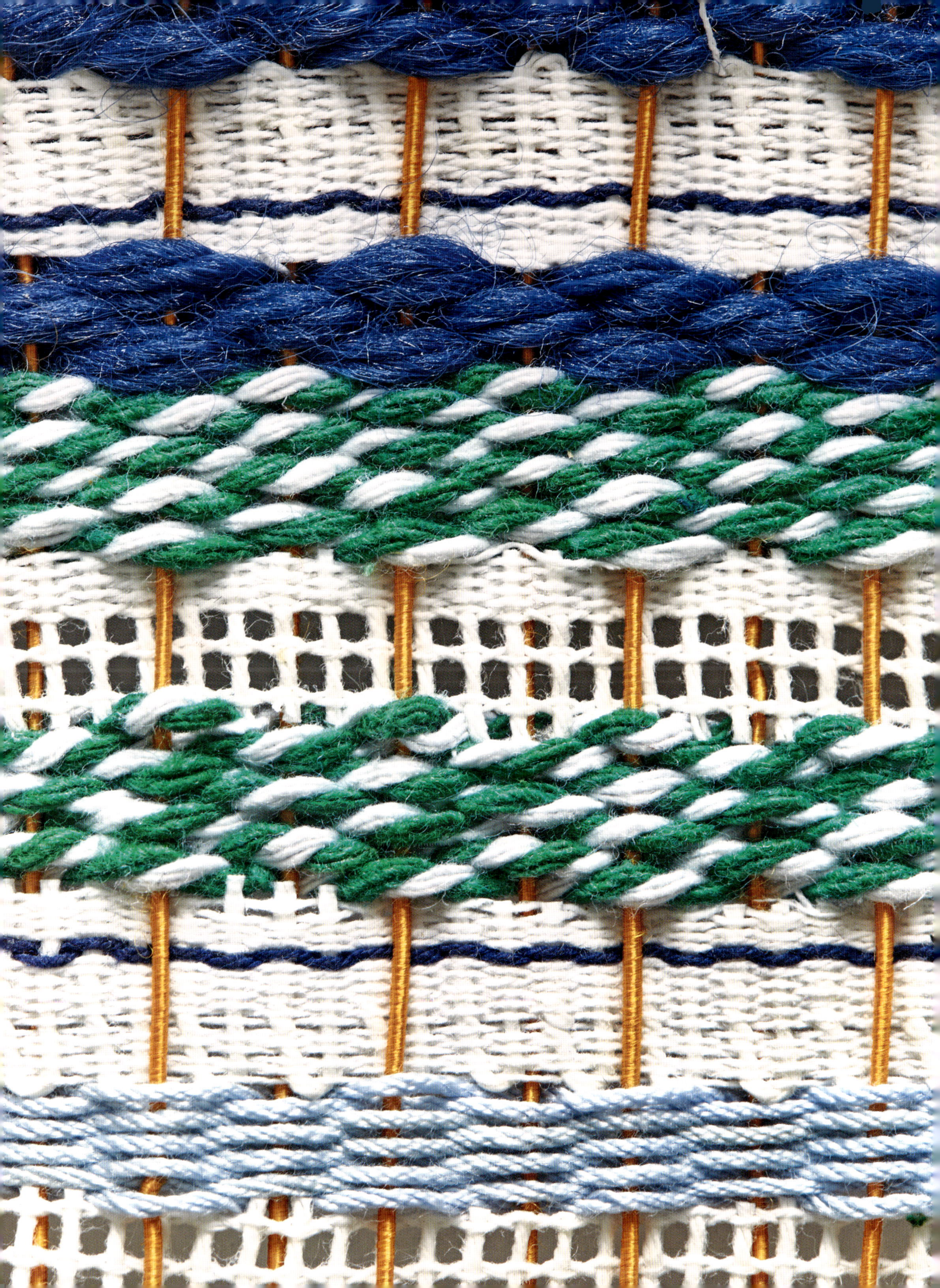

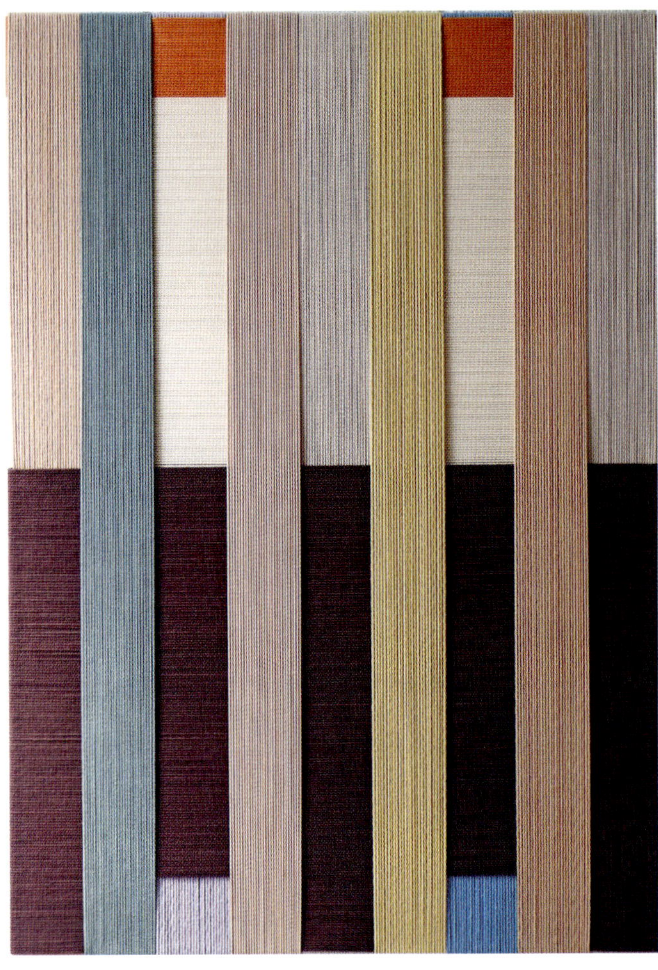

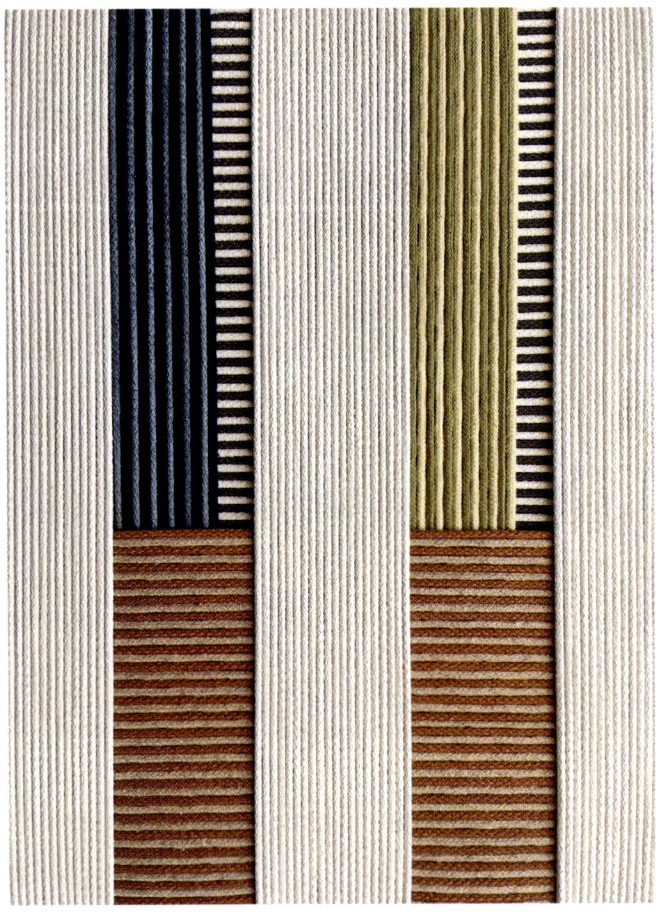

FEATURED ARTIST

Jo Elbourne

Jo Elbourne's distinctive 'wrapping' work necessarily imposes constraints on her pieces, and started with a spool of charity-shop rope, as she explains:

'I didn't have a plan, but I had dining chairs where the seats pushed out, so I started wrapping the rope around the frames and through trial and error ended up with my own take on seat-weaving.

'I was out of work, and unexpectedly in a new town, with time on my hands but not much by way of resources, so everything involved accessible, affordable materials like string and charity shop wool from abandoned knitting projects; things I could try and then undo and re-use. There was immediately something satisfying about the weight and fluidity of the cord and that it's both soft and robust. And there was an element of transformation in creating a taut, rigid surface from something soft through tension alone.

'I have a pared-down process, using one material and a specific technique, working only along parallel and perpendicular lines. Each piece is created in iterative movements, wrapping fine yarn – or braided cord for larger works – row by row to form a taut and layered surface. The act of making is necessarily methodical and ordered, but is offset by a free and intuitive approach to colour and proportion that is not restricted by theory or rules.

'My work deals both literally and figuratively with the notion of managing tension, from the physicality of working with a soft material to maintain a taut structure, to the deeply grounding and restorative tactile benefits of making by hand. The inherent linear nature of the technique lends itself to architectural forms. The postcard-sized works utilise this with abstracted three-dimensional forms, bringing together bold and often disharmonious colours with nods to de Chirico's dreamlike settings and the "emotional architecture" of Luis Barragán.

Above left: Jo Elbourne, *International Colouring Contest*. Wrapped string

Left: Jo Elbourne, *Pillars*. Wrapped string

'The viewer is positioned within tiny imaginary scenes wherein heavily simplified elements of architecture and landscape – such as illuminated doorways, shady corners and objects on distant horizon lines – conjure up ideas around possibility, clarity, uncertainty and far-off futures. Rather than literal representations, the architectural compositions are a vehicle to create illusions of spatial depth, light and shadow, and can just as easily be read as images of abstraction.

'The larger works enter an even more abstract field, focusing on asymmetrical arrangements of shape and colour without narrative or specific influence, with the linear nature of the technique determining the hard edges of the composition. Generally works are made in series, where a moment during the making of one piece will inform the next.'

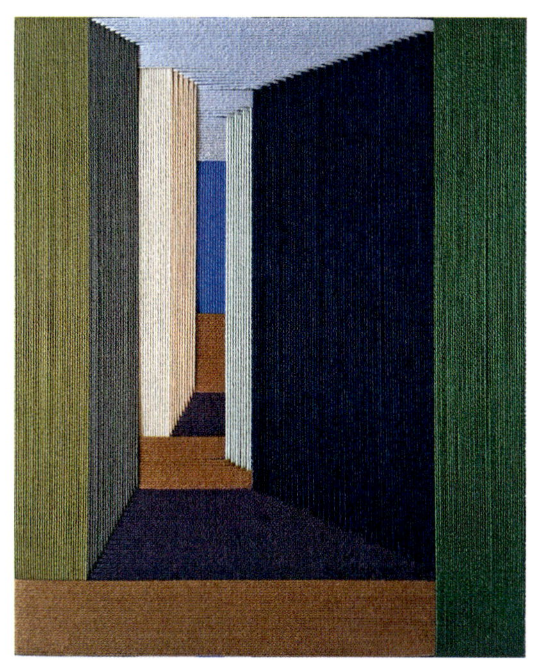

Left: Jo Elbourne, *Keyframe*. Wrapped string

Below: Anne Kelly, *Wrapping Sample and Threads*. Mixed media wrapped piece with threads on card

WORKSHOP

Wrapping threads

A fun wrapping project can be used with a card or wood base.

MATERIALS REQUIRED

- Base card or wood, 15 x 15cm (6 x 6in)
- Heavy perle cotton threads or embroidery cotton (full strands). You can also use strong (reinforced or darning) wool, or thin string in a variety of colours and some neutrals
- Thick needle for punching card or small nails and hammer for wood

METHOD

1 Mark the edges of the card or wood with dots 0.5cm (¼in) apart.

2 Pierce the card or hammer the small nails into place.

3 Make a base 'web' of lines, running across the surface of your piece.

4 Work diagonally next, creating a pattern, until you have built up some depth.

5 Keep wrapping and joining until you are happy with the finished piece.

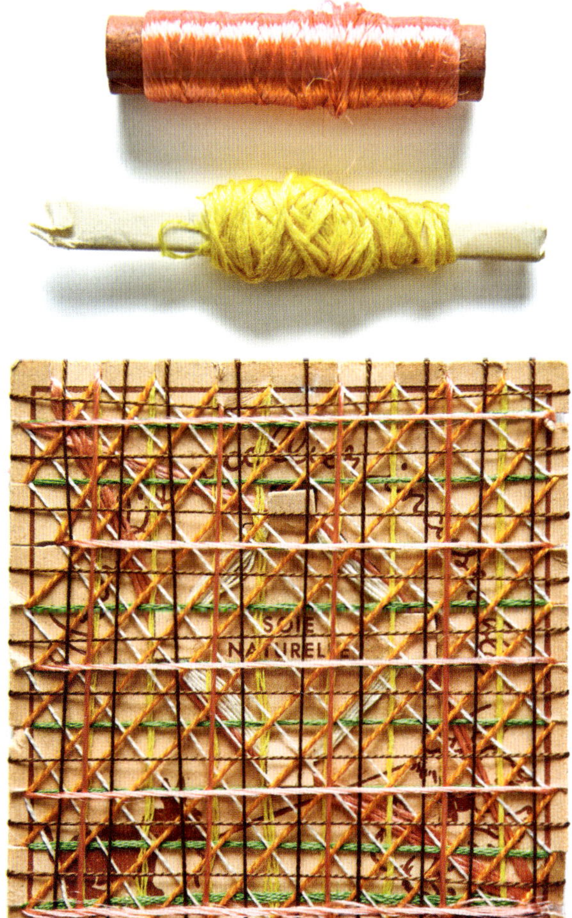

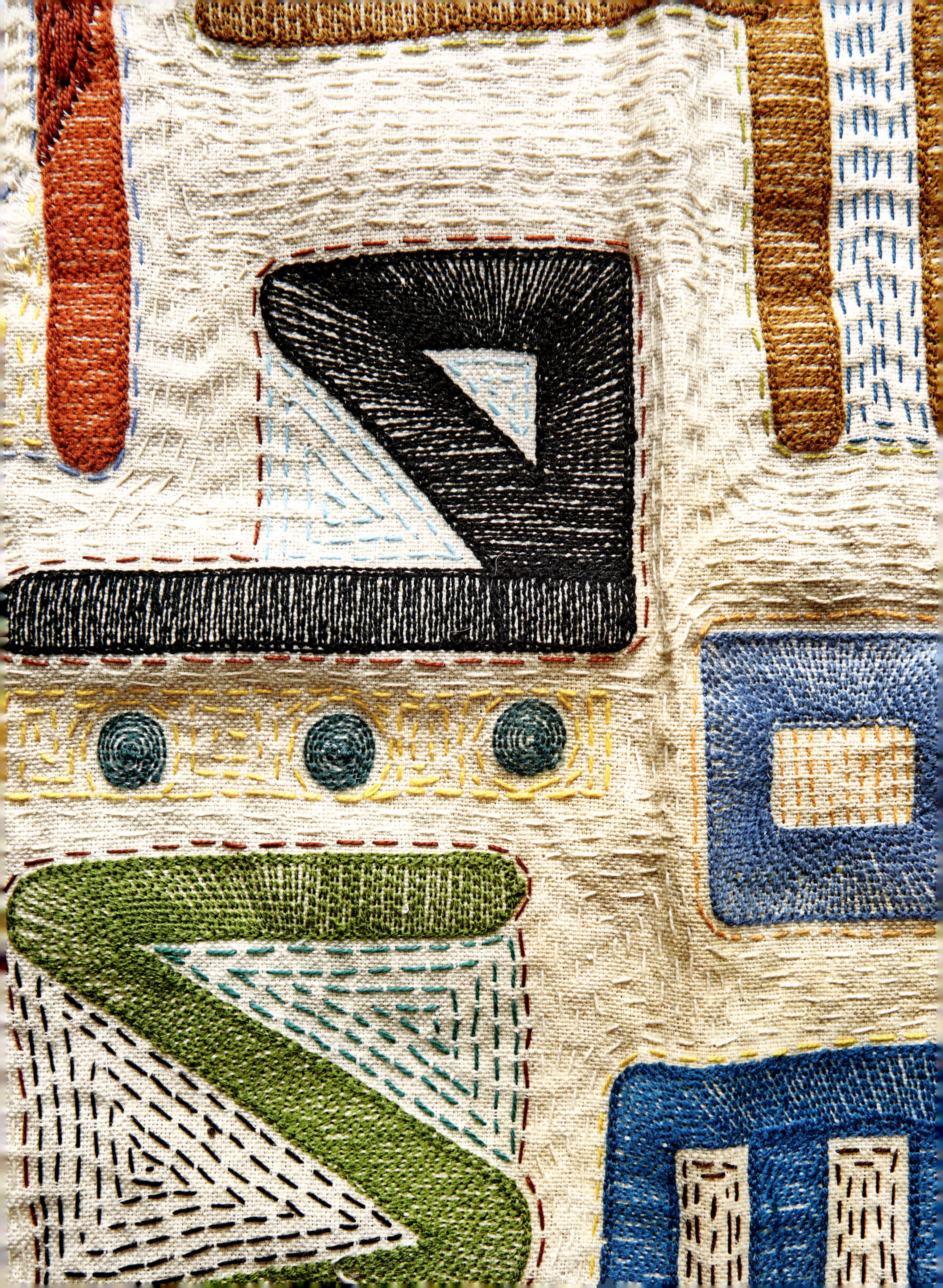

KANTHA AND BORO TECHNIQUES

These techniques stem from the traditional stitching of two great cultures in India and Japan. Although they can appear similar, they had, and have, very different histories and intentions.

Kantha technique

Kantha originated in West Bengal. It was originally used as both a quilting stich and a decorative one – and is still used as both.

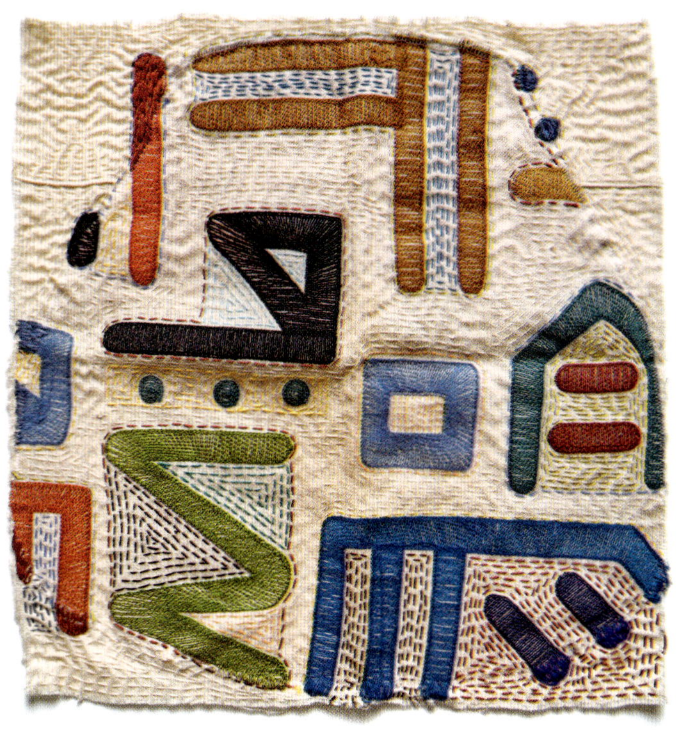

Opposite: Anne Kelly, *Kantha Sample* (detail). Stitched textile

Below: Anne Kelly, *Boro Bird*. Stitched textile

Above right: Anne Kelly, *Kantha Sample*. Stitched textile

Right: Anne Kelly, *Boro Bird* (detail). Stitched textile

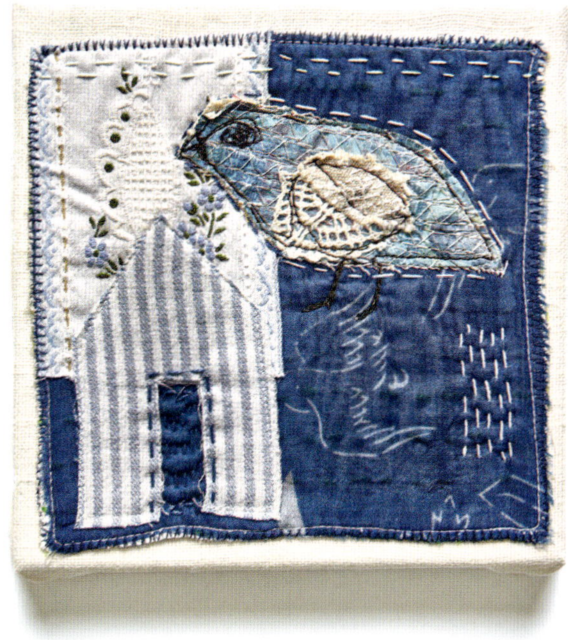

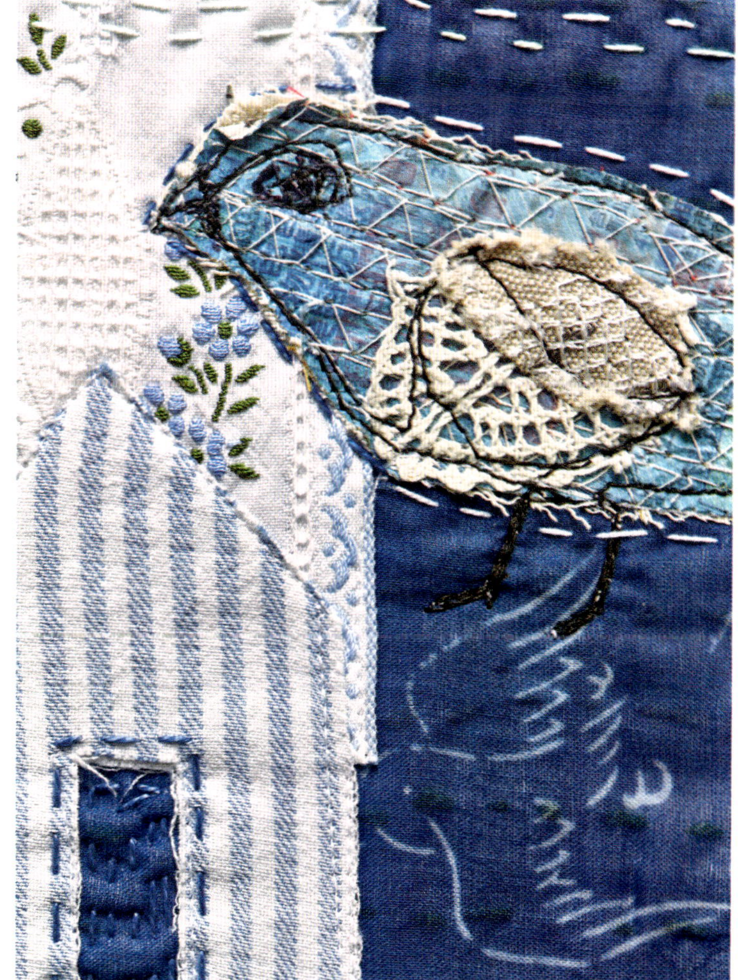

Boro technique

Boro was traditionally used as a repairing stitch, and has morphed into a decorative one. For our purposes, this technique can be useful both for creating texture and areas of shape and shading in stitched work.

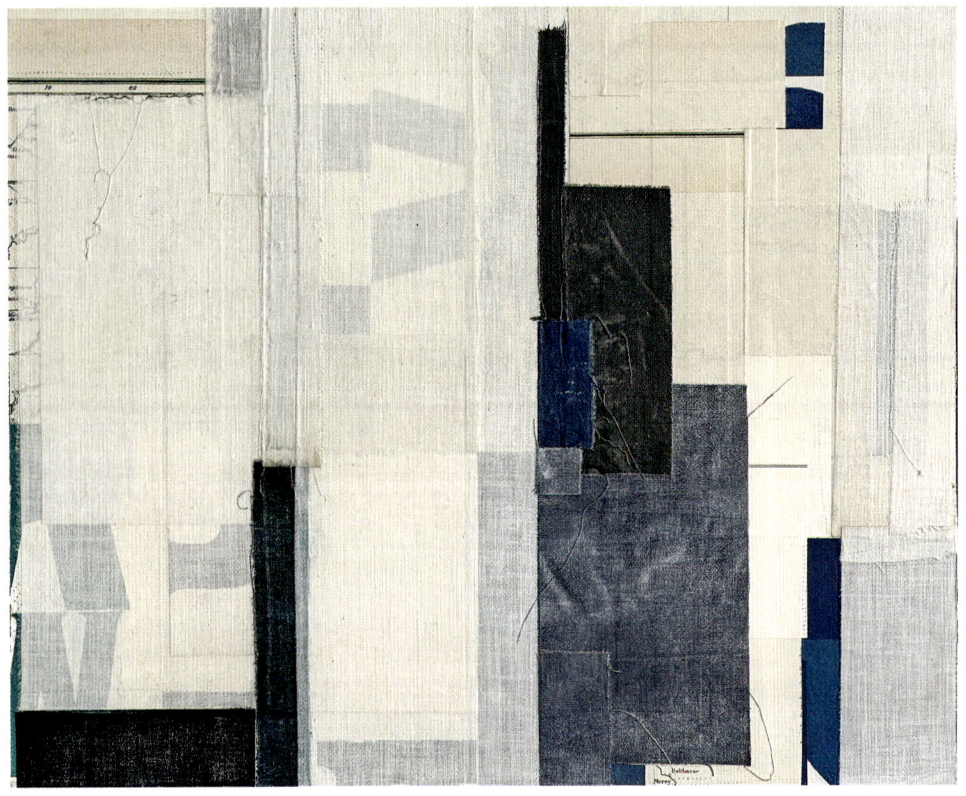

Abstract Boro

Here are some samples of Boro being used in an abstract format. Sarah Z. Short is a mixed media collage artist working in paper and cloth. We take a closer look at her work in Chapter Six.

Above: Sarah Z. Short, *Balthazar*. Mixed media textile collage

Below left: Sarah Z. Short, *Purple Star*. Mixed media textile collage

Below right: Sarah Z. Short, *Droplet*. Mixed media textile collage

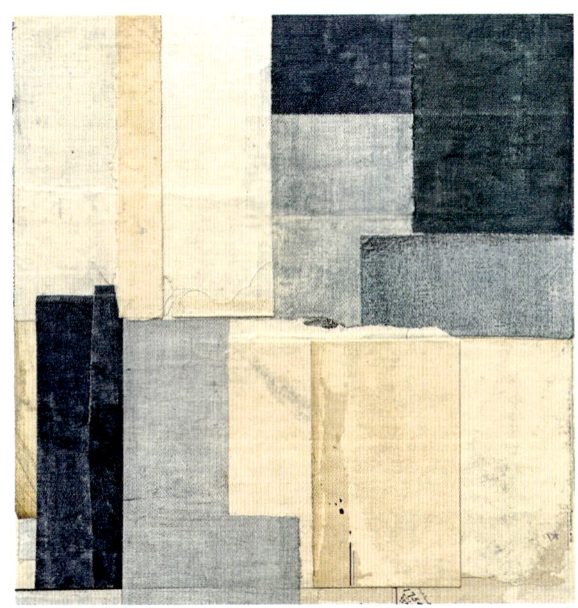

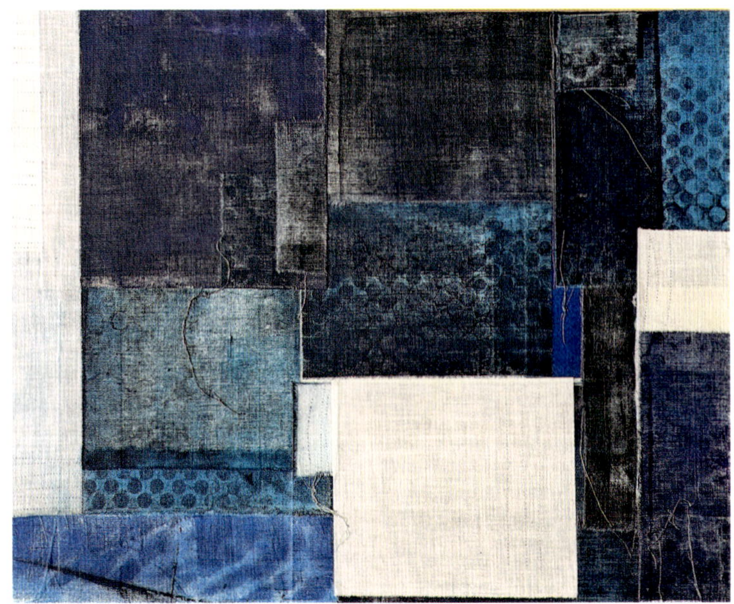

A MID-CENTURY KANTHA JACKET

I found this jacket on a visit to India. It appealed to me as it is covered in Kantha stitching, covering the cloth from which it is made. I decided to add a 'placket' or decoration around the opening, using mid-century textiles and embroidery. I embroidered using perle cotton to reflect the colours on the jacket.

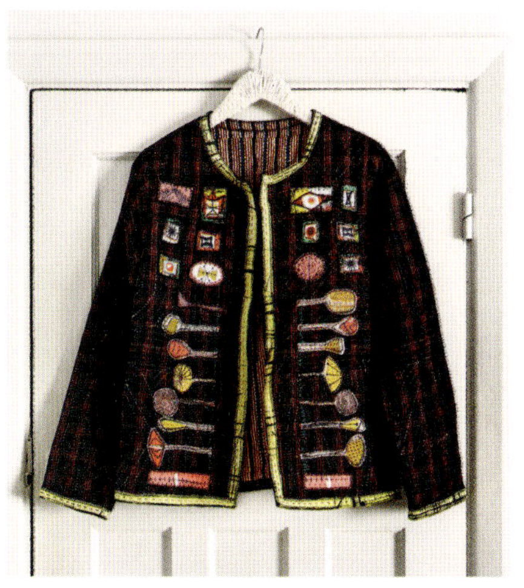

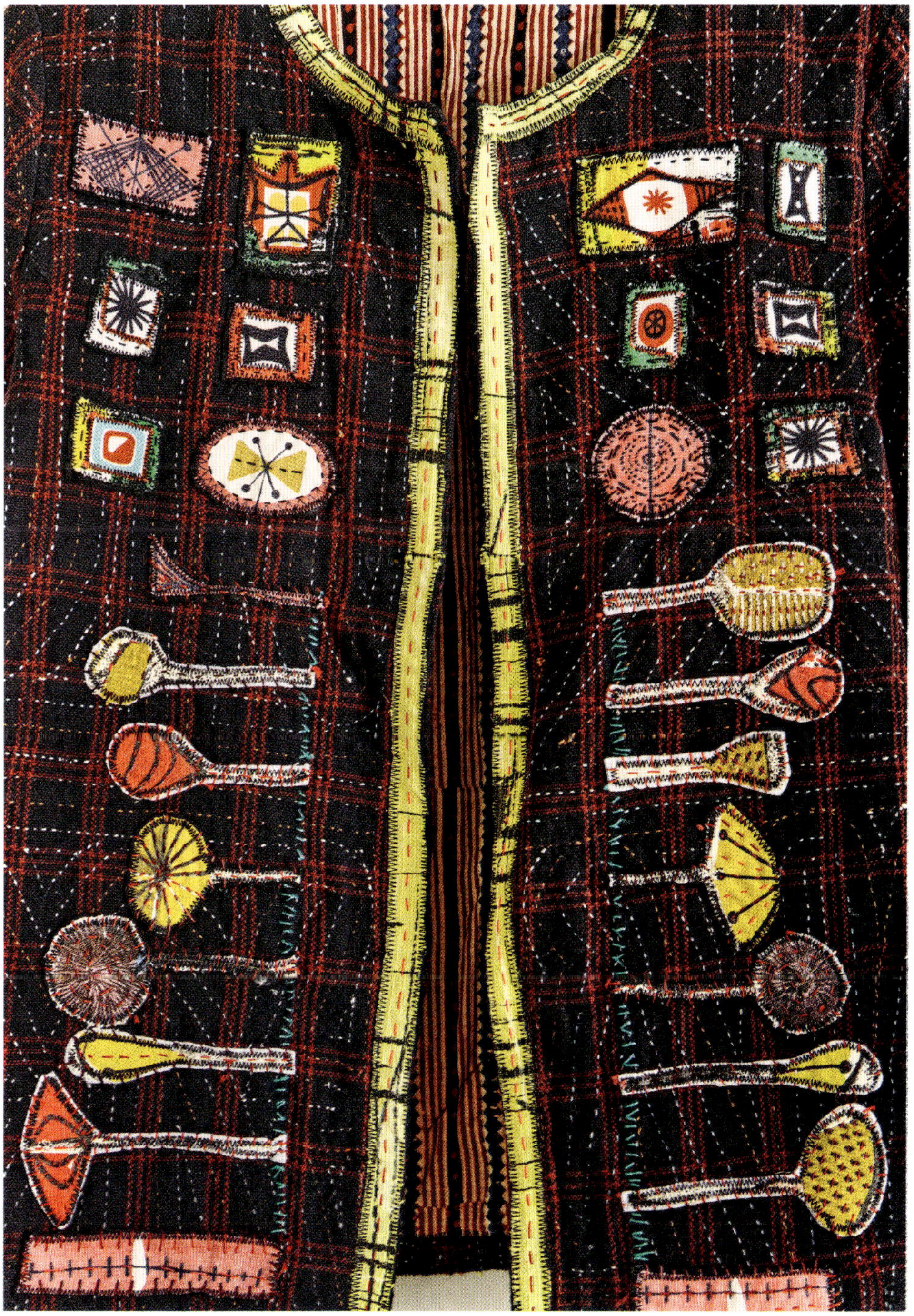

Above right: Anne Kelly, *Mid-century Meadow Jacket*. Stitched and appliquéd jacket

Right: Anne Kelly, *Mid-century Meadow Jacket* (detail). Stitched and appliquéd jacket

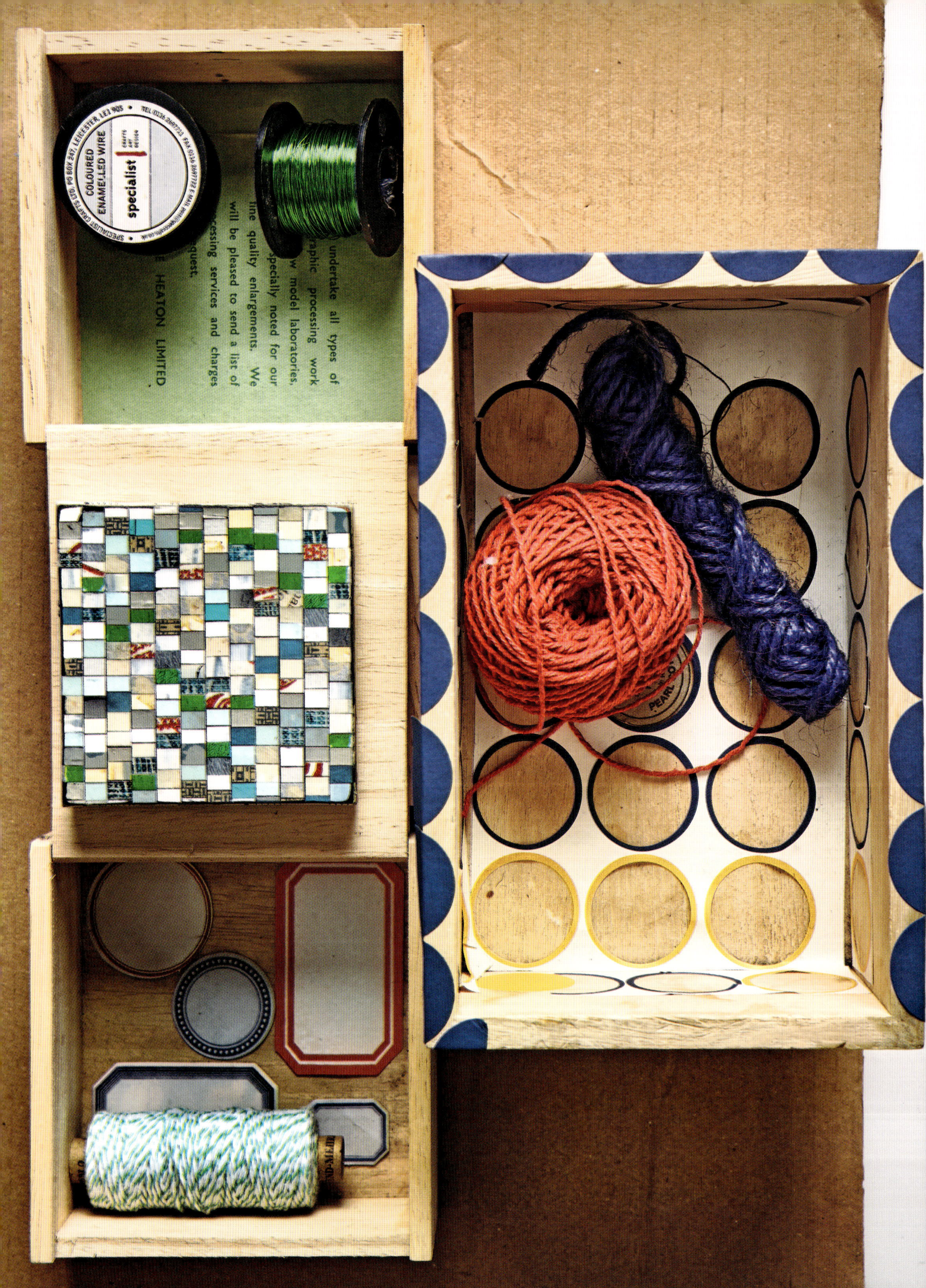

COLOURED ENAMELLED WIRE
specialist
undertake all types of
raphic processing work
w model laboratories.
pecially noted for our
fine quality enlargements. We
will be pleased to send a list of
ocessing services and charges
quest.
HEATON LIMITED
PEARL

Chapter 5

CAPTURING MEANING

'The truly modern artist is aware of abstraction in an emotion of beauty.'

Piet Mondrian (1872–1944)

Opposite: Anne Kelly, *Inspiration Box, Materials*. Mixed media in wooden boxes
Sarah West, *Untitled Canvas*. Mixed media

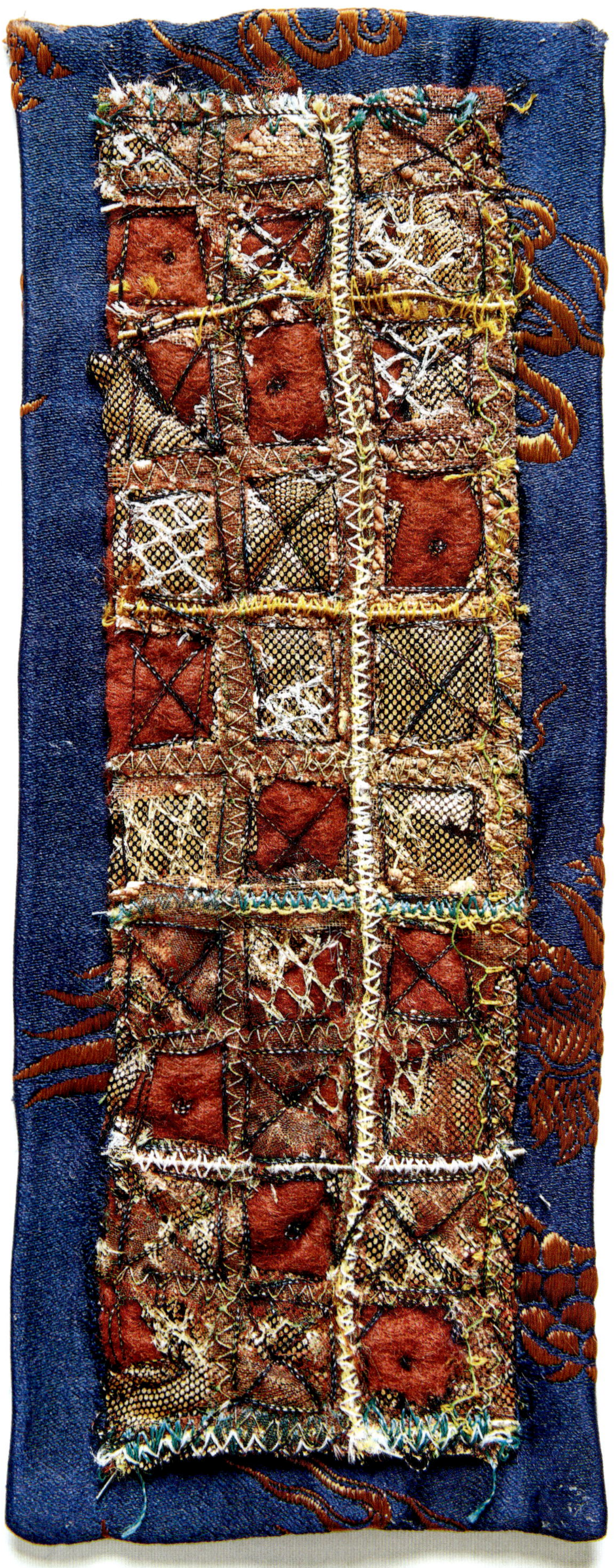

It is a challenge to infuse abstract work with emotion and beauty, but it is a worthy one. Although abstract work is more interpretative in the viewing, it can also be more atmospheric and evoke memories and emotion. In this chapter we look at ways to imbue your work with a deeper meaning and feature artists who work in this way.

Abstract work can be viewed subjectively, but the more it is studied and viewed, the more possibilities open up. As we have seen, colour, line and shape are integral to making abstract work. Writing, reference and media take it to another level. It is wonderful to see this kind of work in situ and spend some time contemplating it.

Left: Anne Kelly, *Abstract Sample*. Stitched sample

Opposite: Anne Kelly, *Wrapping Sample* (detail). Mixed media wrapped piece

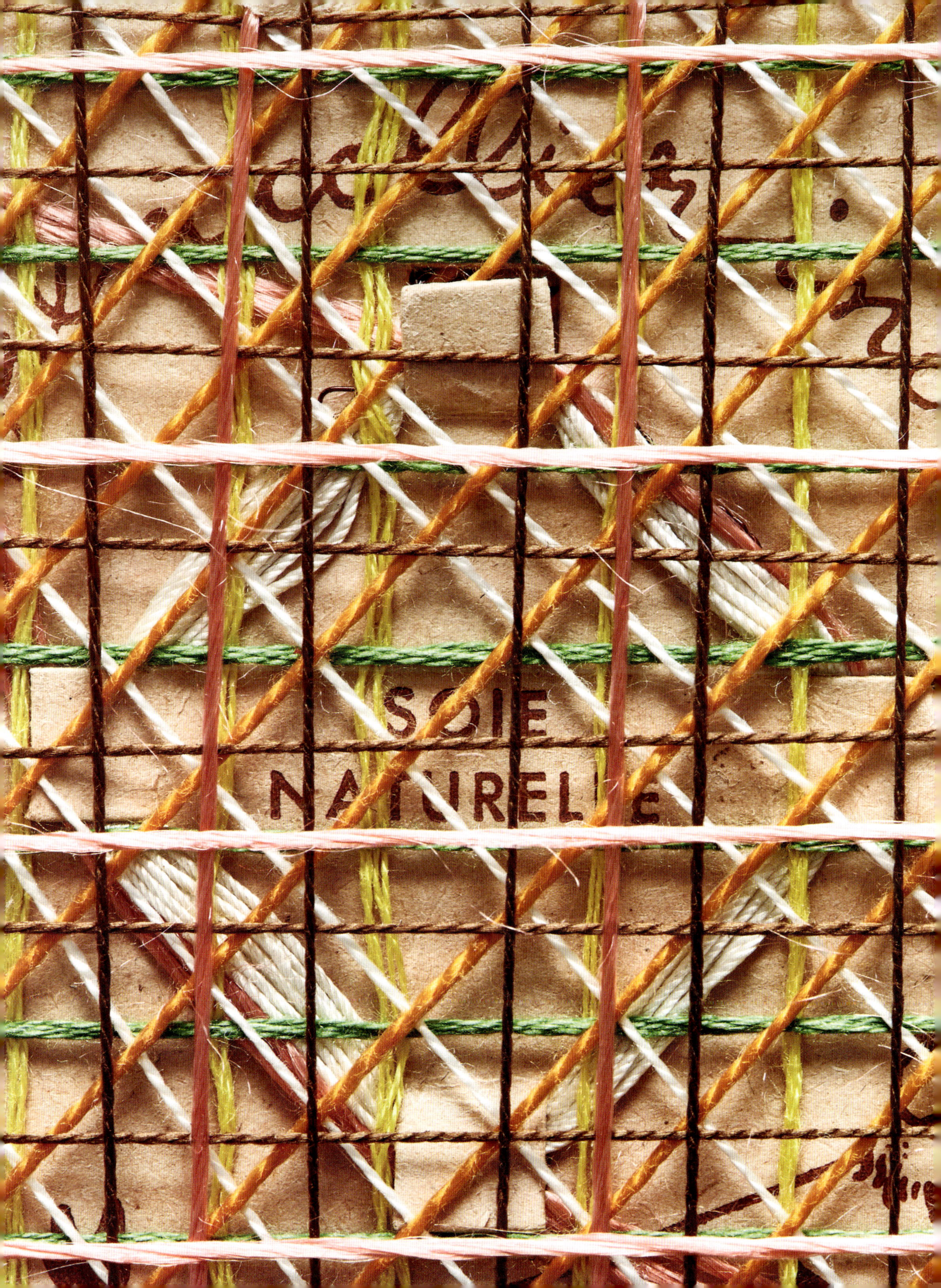
SOIE
NATURELLE

MEMORY

I had a short stay in hospital in early 2024 and decided to keep a stitch journal while I was there. I had assembled the 'pages' before my visit and was able to add stitch and some appliqué with a small bag of threads and fabric. The *Hospital Book* is a record of my thoughts and emotions at the time. It was very comforting in that environment to return to stitching and making when possible. When I returned home, I added some printed elements and completed the stitching.

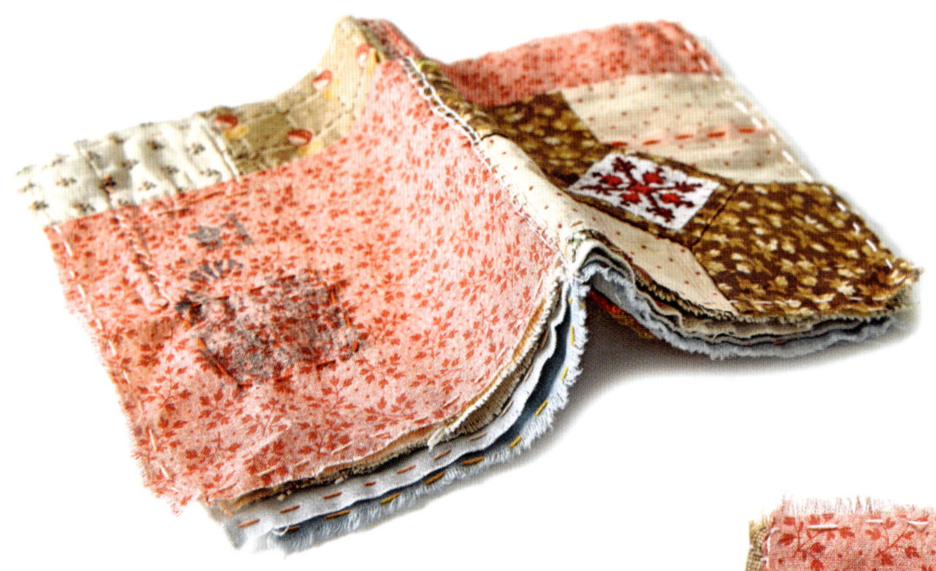

Above & right: Anne Kelly, *Hospital Book* (covers). Mixed media textile

Opposite: Anne Kelly, *Hospital Book* (pages). Mixed media textile

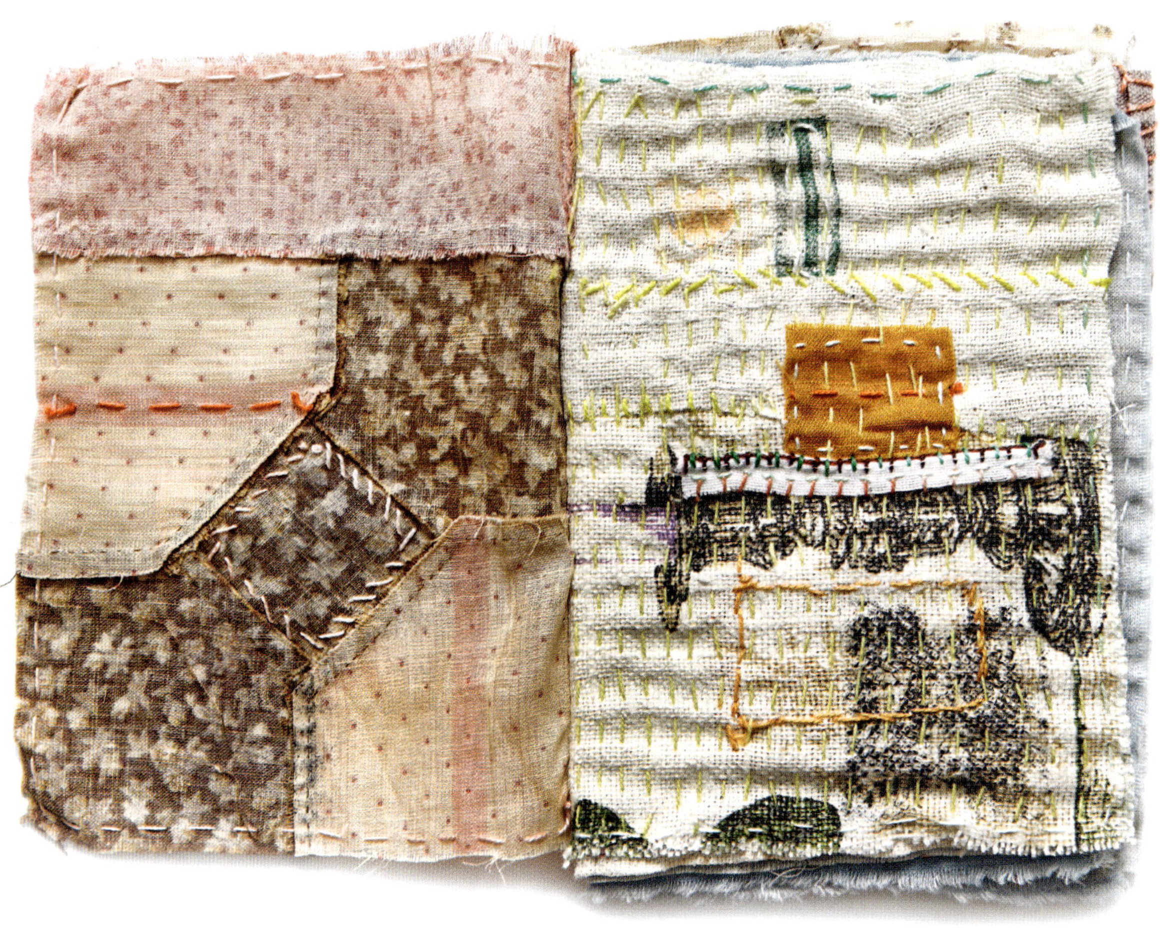

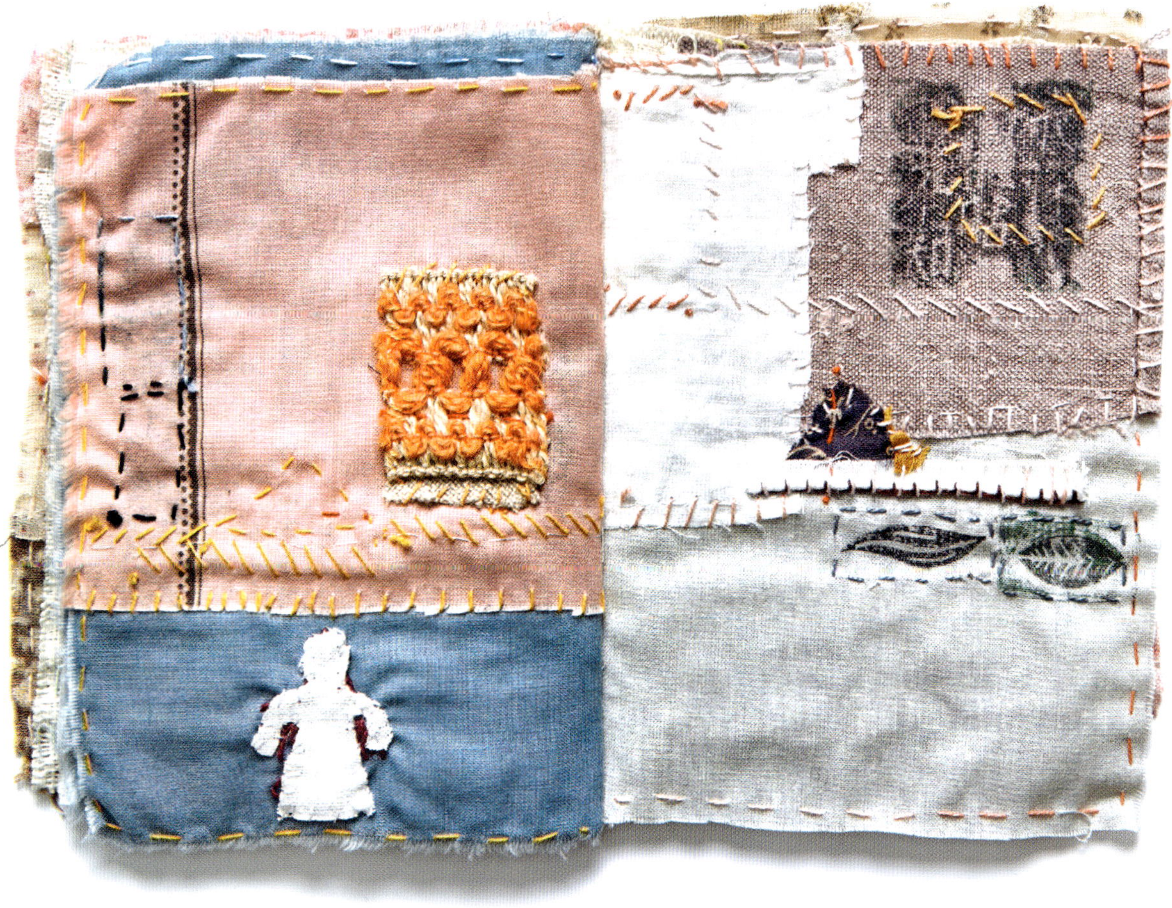

FEATURED ARTIST

Alysn Midgelow-Marsden

Alysn's work bridges a gap between intention and interpretation. Her sensitive markings leave space for the viewer to add their thoughts, as she says:

'Once upon a time, I trained as an embroiderer. Though always combining traditional with experimental skills, it is now more accurate to describe myself as a sculptor using some textile and stitch techniques. This removes the stereotypes and expectations associated with the term "embroiderer" and leaves me freer to create works that fluidly combine concepts and aesthetics.

'I create pieces that have both physical and psychological articulations and movements that are contemplative and expressive, often layered with ideas and intricate detail. It is also important to me that there is space in the work for a viewer to imbue it with their experience and understanding. By removing recognisable objects or obvious narratives I can open my works to be personally meaningful to them.

'My process in a nutshell is to begin by "taking leap after leap in the dark" (Agnes de Mille) and "stop in interesting places" (Paul Gardner). Although the works rarely have representational form, many start with inspiration from a visual spur and are developed through a combination of observation, drawing, material explorations and writing. At some point I will define the key components that need to be expressed in the final piece, as this helps me to hold myself accountable to my core vision.

'During the development of work I will flit between the concept and the hands-on experience of making. It is through these that the form and nature of the piece will resolve itself. This improvisational approach can of course be unsettling. The unpredictability can mean that time is spent on works that are never completed, but I have learnt to trust that flowing with the process will lead me and it overrides any concerns and takes me to more interesting places.

'The techniques I utilise are dictated by the needs of the piece or collection and I remain open to traditional skills as much as innovative ones, from stitch and crochet to digital techniques, welding and so on.'

Below: Alysn Midgelow-Marsden, *Limited Longing*. Mixed media textile

Opposite (above): Alysn Midgelow-Marsden, *The Possibility of Perfection*. Mixed media textile

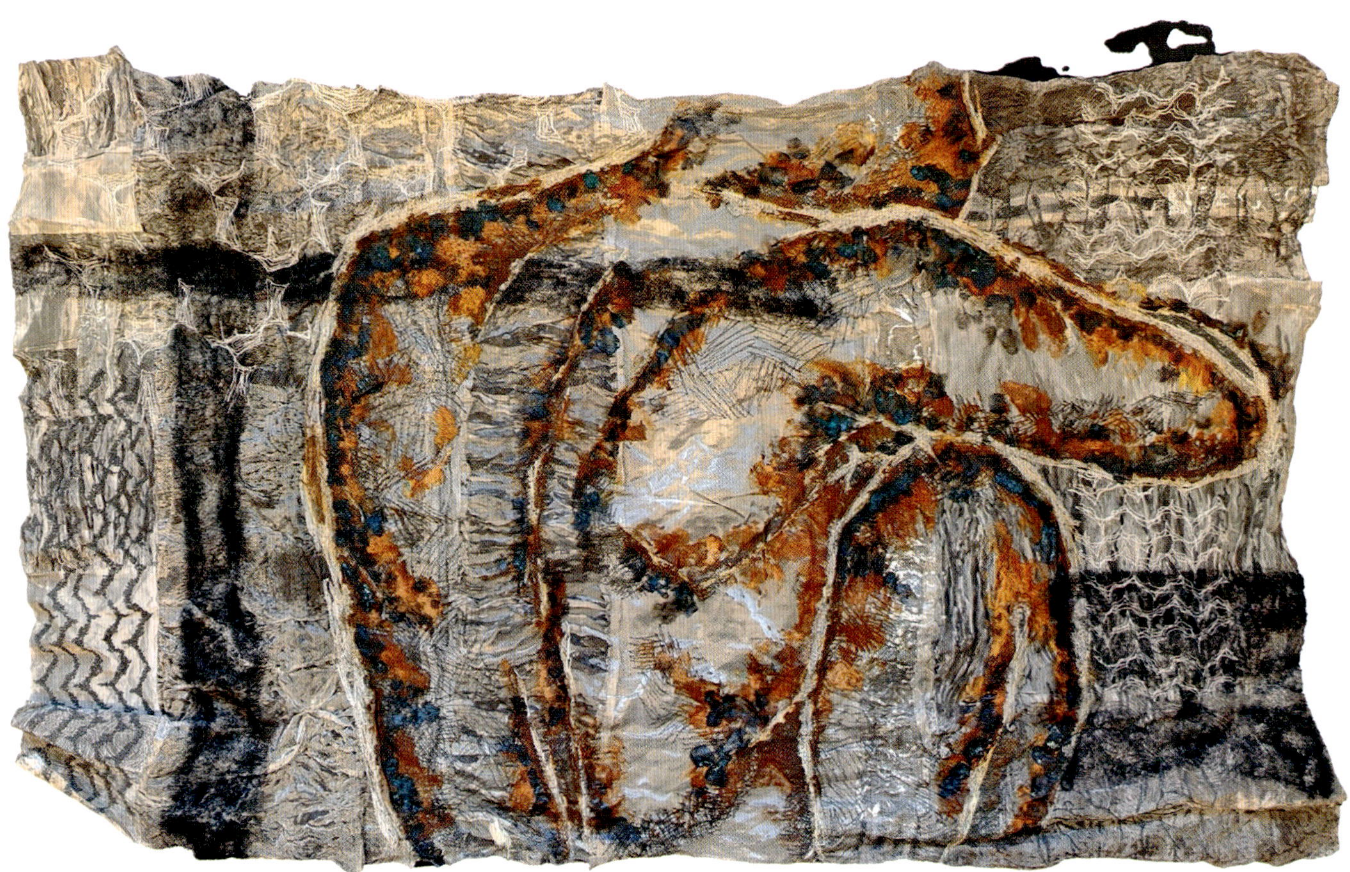

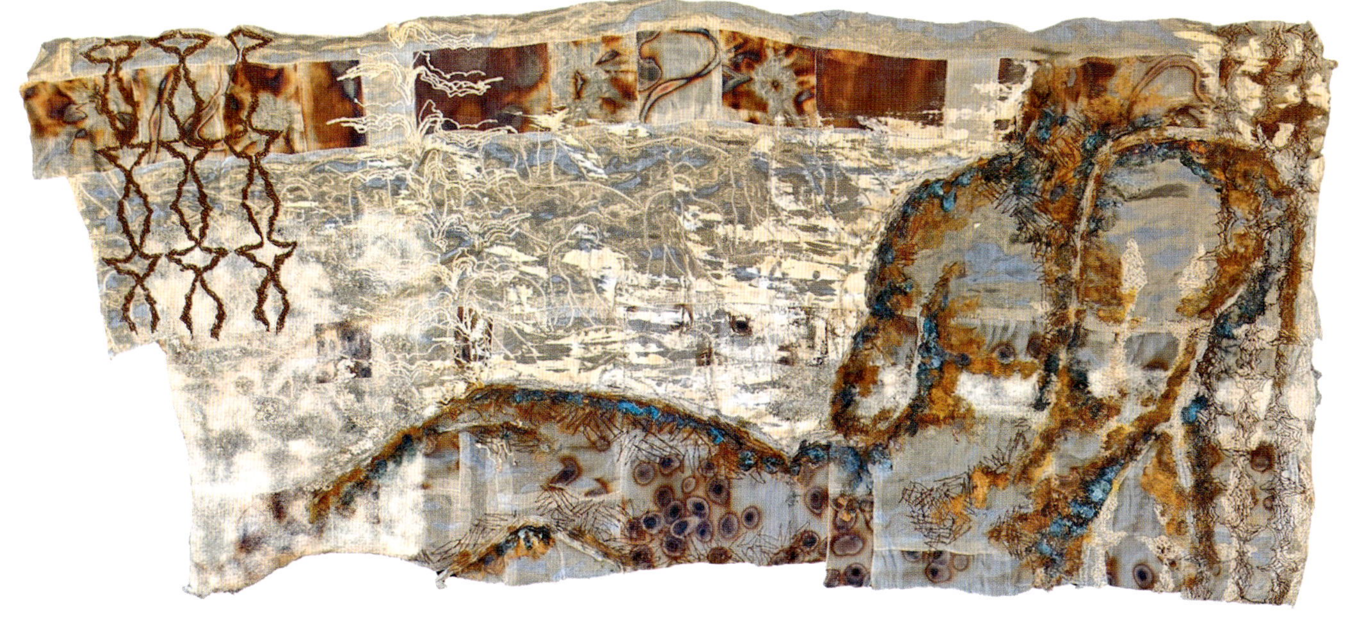

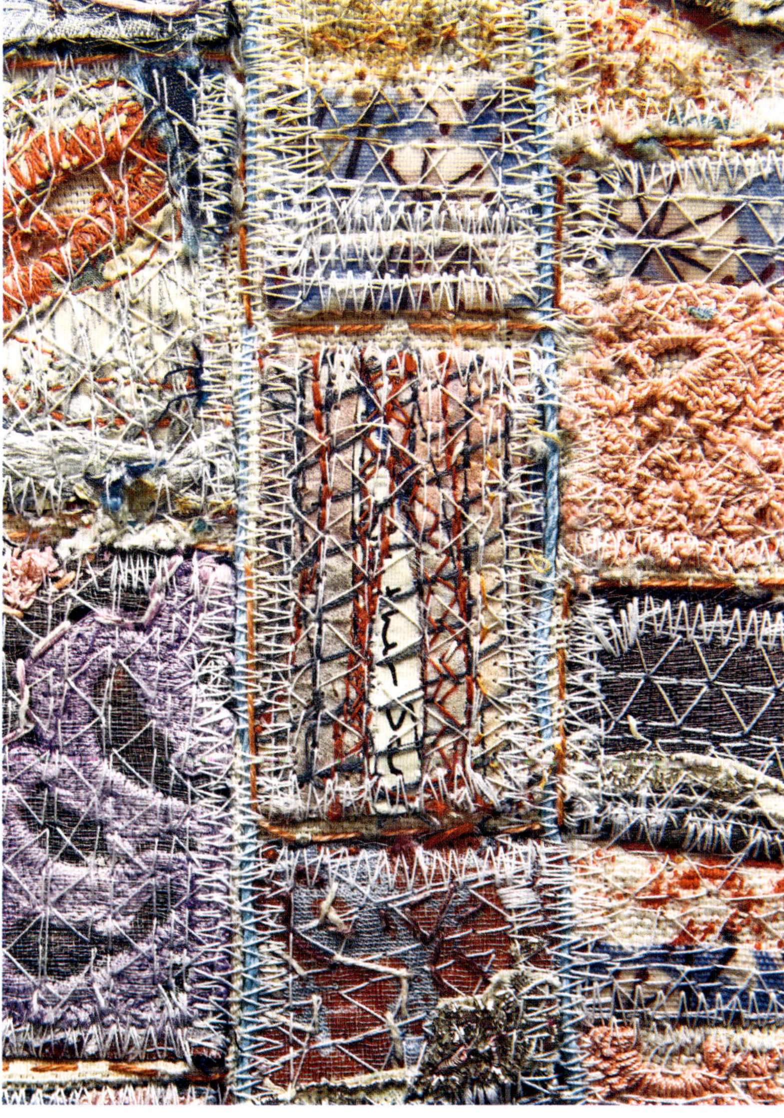

WORDS

Adding words to a piece of work can ground it and expand on your own thoughts, making it more accessible to the viewer. Titles, likewise, can make you think about what the piece means to you and its purpose. We will be thinking about how to add words to your work and where to place them.

Left: Anne Kelly, *Cell Bag* (detail). Mixed media, printed and patched textile

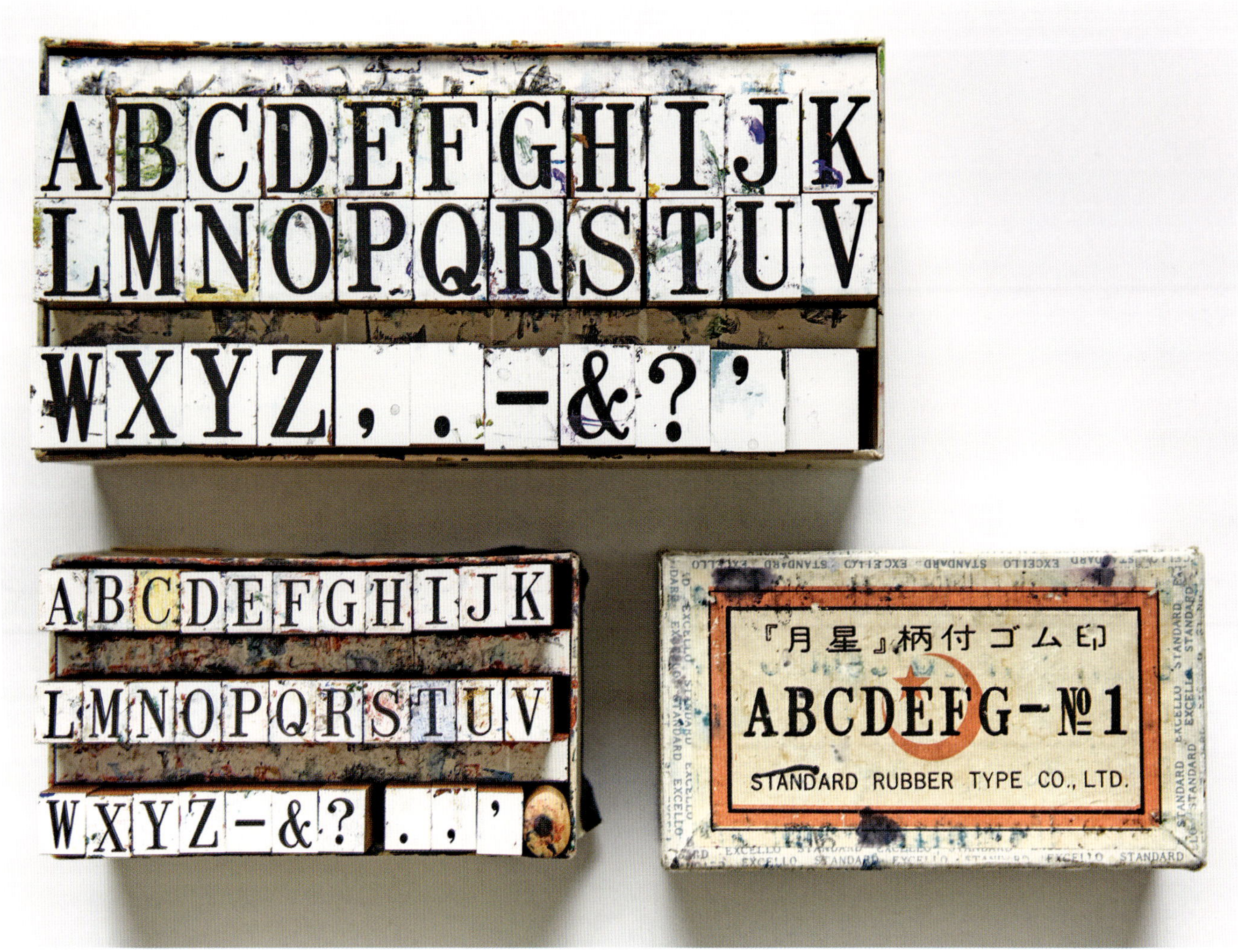

ADDING WORDS TO TEXTILE ART

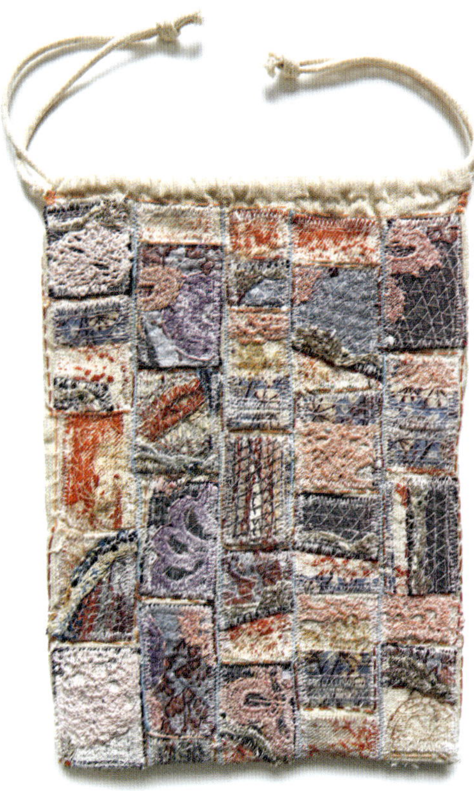

MATERIALS REQUIRED **(DEPENDING ON TECHNIQUE)**

- **Rubber stamps** – fabric ink pad or fabric paint, small sponges and palette
- **Hand stitching** – embroidery thread, hoop and needles
- **Machine embroidery** – free-motion foot, machine embroidery thread
- **Image/word transfer** – printer-friendly fabric sheets or T-shirt transfers

All of the above techniques are useful for adding words. I like to write by hand the phrase or words to be added on a piece of baking or tracing paper and then place it over the area where I am going to add it, to check the size and position of the writing.

I suggest printing and stitching the writing onto a separate piece of fabric and then adding it to your piece. If you use an organza or thin cotton, the writing will merge beautifully into your work.

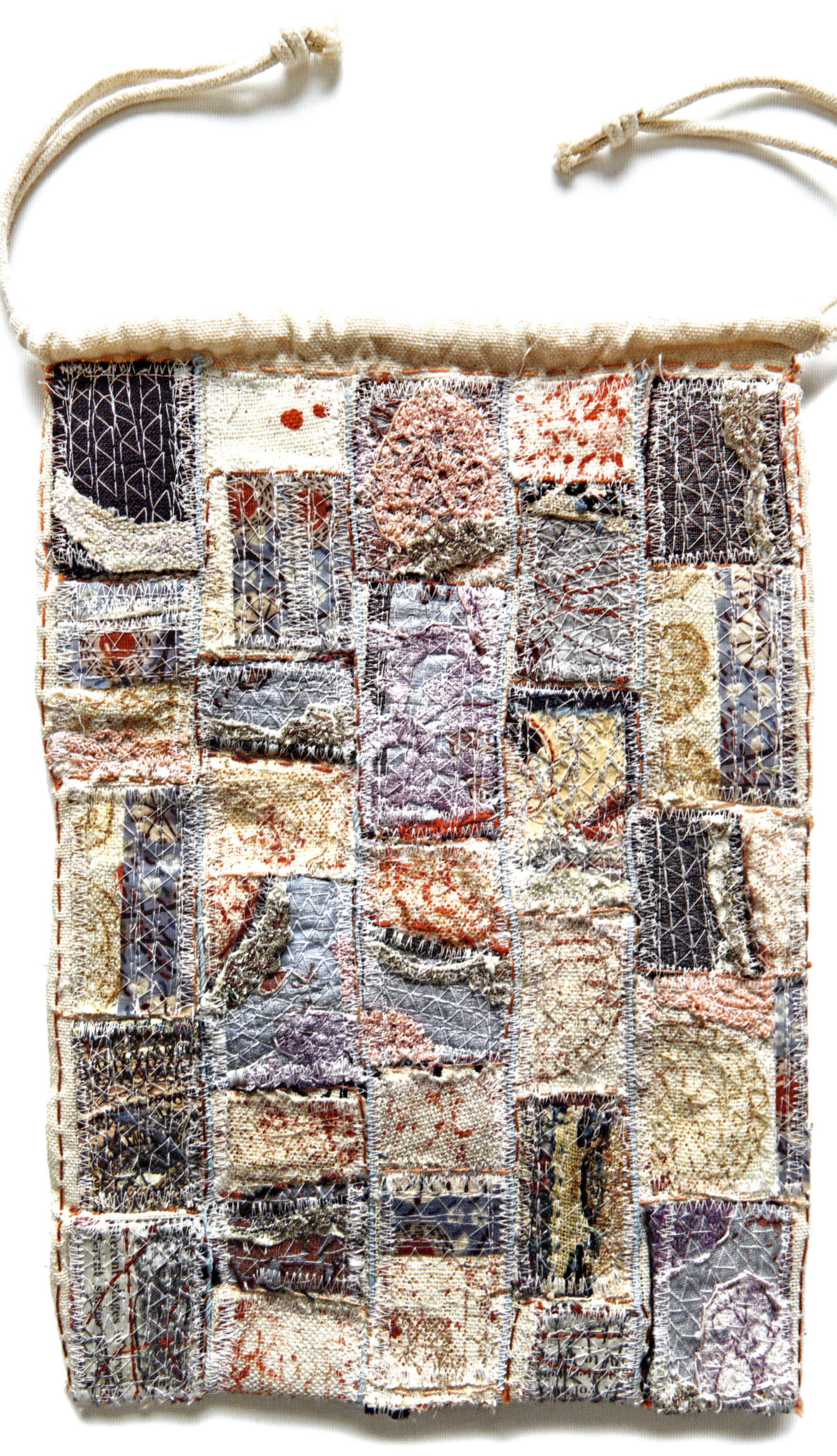

Opposite top: Japanese rubber stamp sets, author's collection

Opposite bottom: Anne Kelly, *Cell Bag* (reverse). Mixed media, printed and patched textile

Left: Anne Kelly, *Cell Bag* (front). Mixed media, printed and patched textile

IMMERSIVE EXPERIENCES

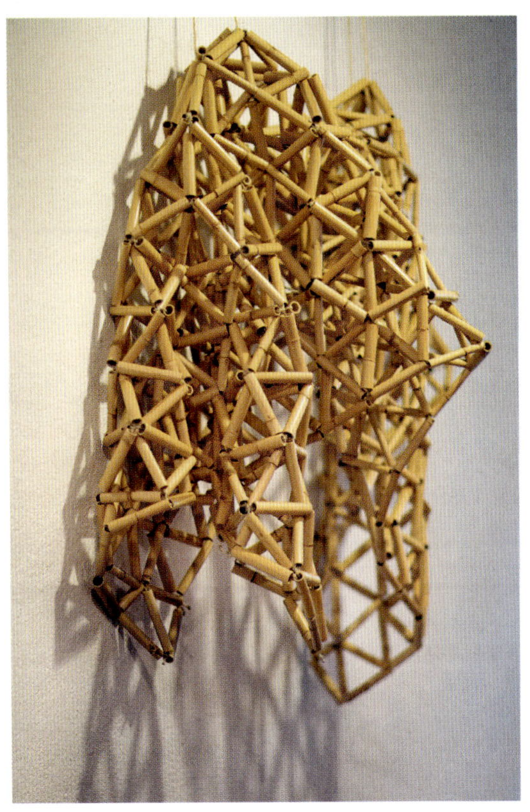

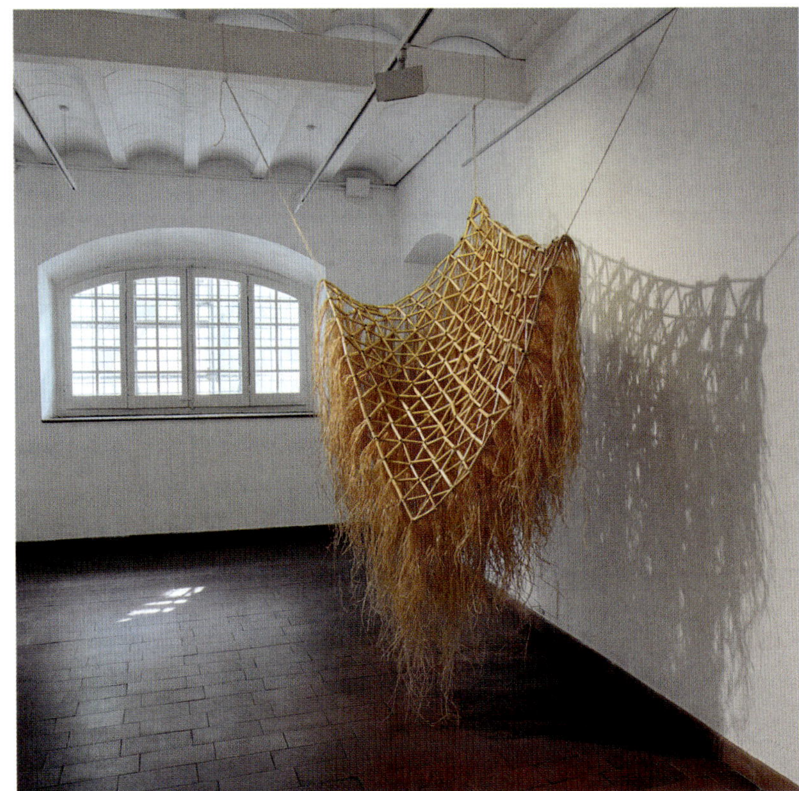

FEATURED ARTIST

Tim Johnson

I met Tim Johnson in Australia while teaching and exhibiting there. His work is informed by an extensive knowledge of basketry. Tim uses it in a myriad of ways to create something special, as he explains:

'As a child in the 1970s I often watched my mother as she laid out paper dress patterns on the floor in preparation for making a whole variety of garments. She was both a seamstress and a prodigious knitter, providing for many of my childhood clothing needs. This seemingly magical transformation from flat fabrics and yarns to wearable clothes fascinated me and doubtless influenced the pathway of my making over the years.

'The simple idea of folding two-dimensional cloth to create three-dimensional forms continues to inspire my work today. I combine it with the familiarity of my preferred choices of natural materials and a variety of traditional basketmaking and textile techniques to provide a wide palette for my creative expression.

'Traditional rain capes from Japan and Spain alongside knotted carpets and thatched roofing have inspired the use of pile in my work. This layering of fibres creates a sense of volume and protection and satisfies my desire to present materials in an uncomplicated way.

'Making stuff takes time – gathering, rummaging, collating and selecting – this may be material grown, harvested or salvaged – from field and wetland, from urban fringe or seashore. It may also be ideas – sifting through experience and teachings, ideas thought and discarded, later to resurface with unexpected connections. Creativity, it seems, takes even longer – hard graft and sore fingers lead to knowledge and occasionally a worthy idea for further exploration.'

Adapted from exhibition catalogue *East Weaves West: Basketry from Japan and Britain*, 2007

Above left: Tim Johnson, *Mountainscape*. Canya/cane (*Arundo donax*), hemp string

Above right: Tim Johnson, *Falling Fringe*. Canya/cane (*Arundo donax*), hemp string and raffia

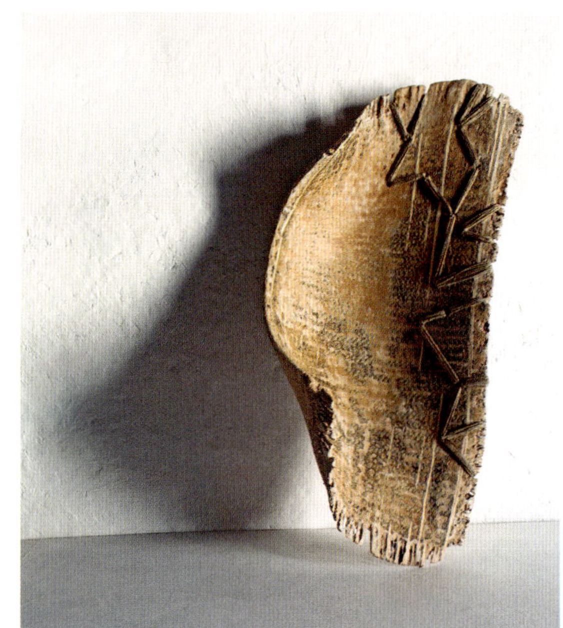

Left: Tim Johnson, *Stitched Curve*. Willow, earth pigments, natural and tarred hemp twine

Below left: Tim Johnson, *Stitched Curve* (detail). Willow, earth pigments, natural and tarred hemp twine

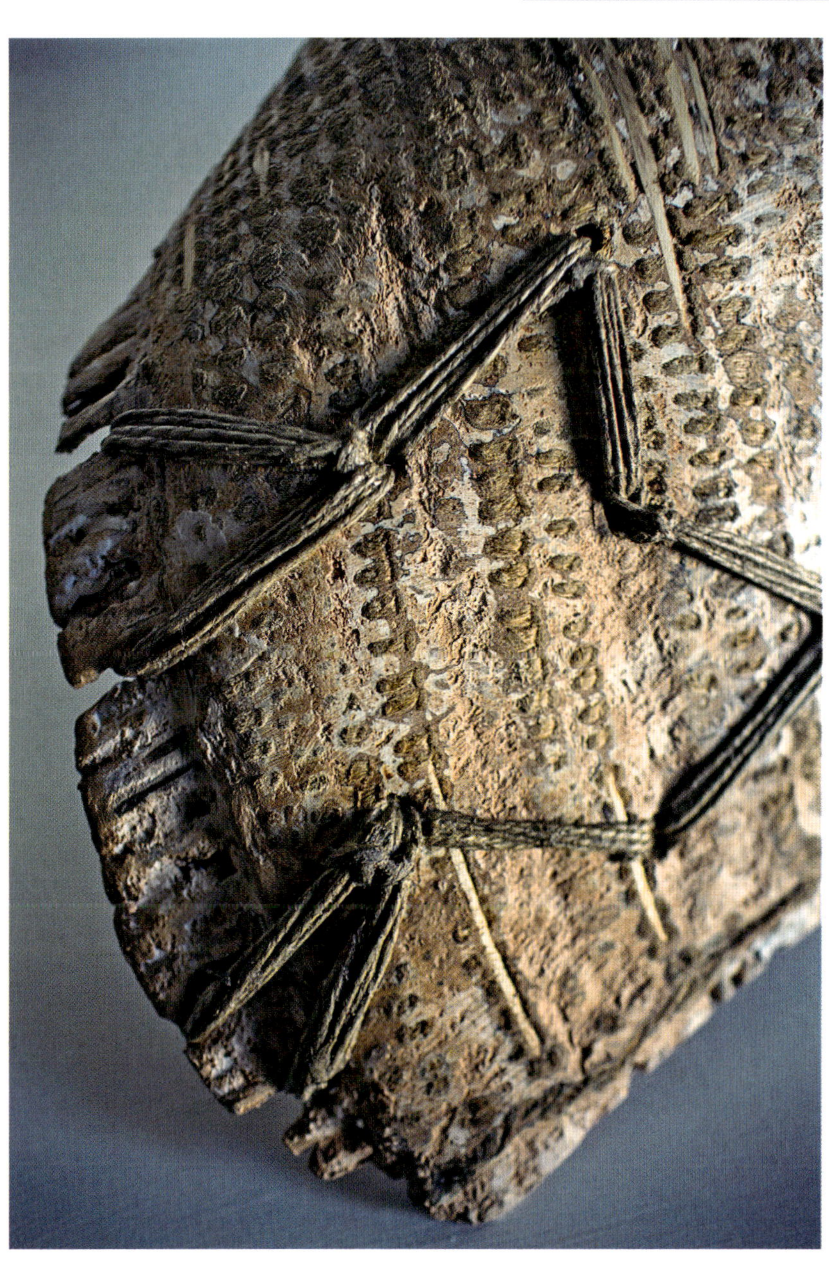

FEATURED ARTIST

Morii Designs

Morii Designs are a contemporary design studio based in Gujarat. They use traditional textile techniques and craftspeople to create textiles for interiors and decoration. The work that they produce has a strong element of repetition and of contemplation, and is well suited to a modern, minimal aesthetic.

The company describes itself as 'an artistic interpretation of age-old legacies; an erudite conversation in a language fading away; a manifestation of global sensibilities rooted in Indian traditions'.

Morii Designs use Jat embroidery (counted thread work) with tiny cross stitches to create geometric patterns that are both visually striking and symbolic. Bela block printing is another technique used by the company, and they work with a family who have used it for one hundred and fifty years. Sujni is another stitch technique the company's designs feature; it uses running stitches to create a dense pattern, similar to Kantha but more condensed. They also use local woven cotton (Kala), grown organically and hand-woven by the local Vankar community.

The overall impression of their work is that it represents the history of traditional Indian craft without being fussy. By its abstract and patterned nature it creates a hopeful and progressive sensibility that makes it stand out artistically. The community support elements of the practice give the work an added depth and sustainability.

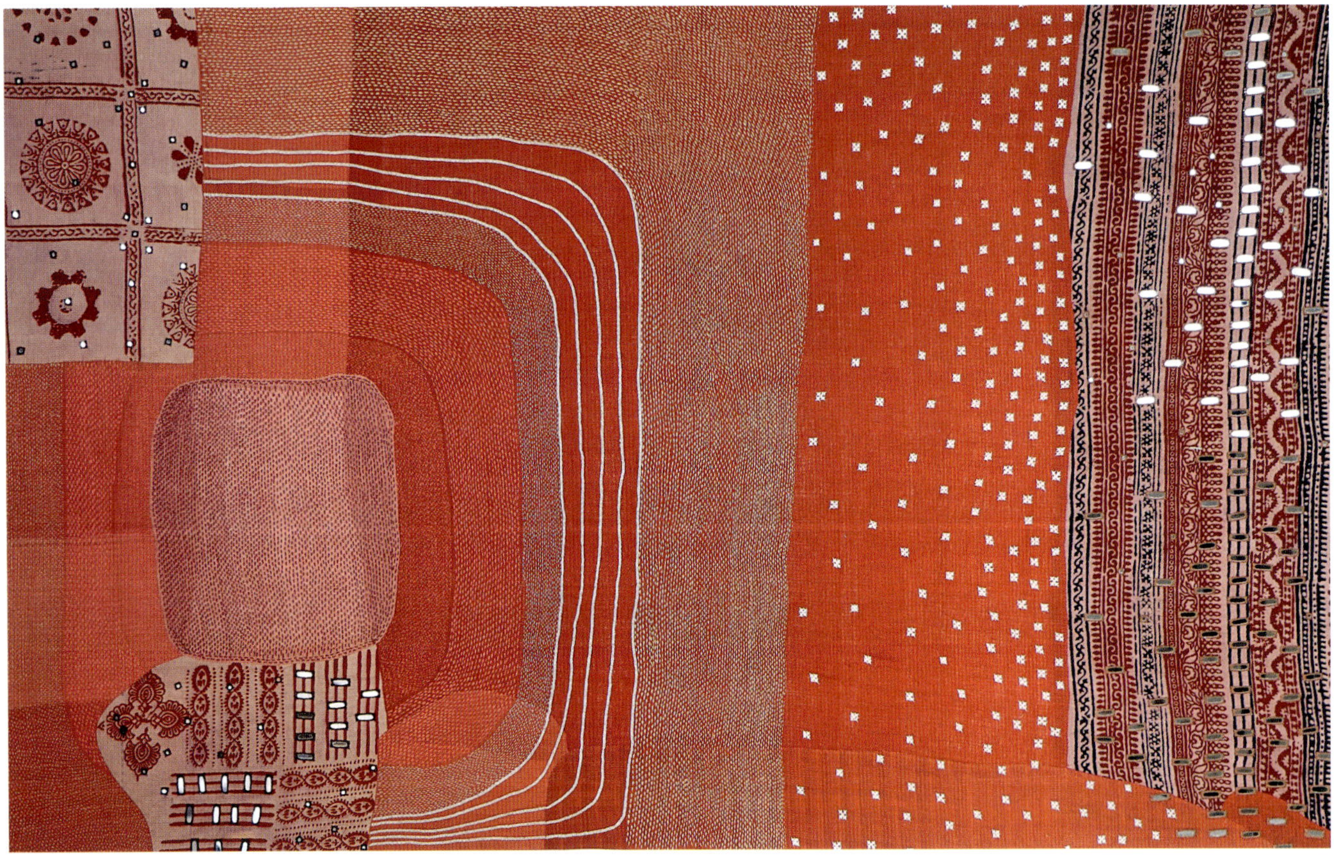

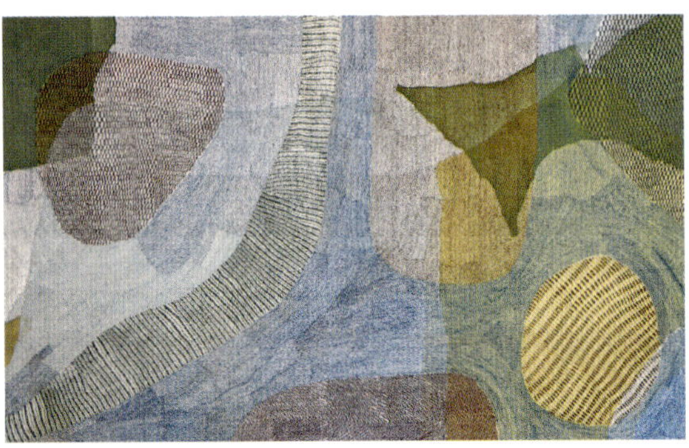

Above: Morii Designs, *Gramya*. Contemporary Rabari embroidery on cloth

Left: Morii Designs, *Advait*. Mixed media textile

TRAVEL SCROLL

Making a scroll piece can be a deeply satisfying way to explore abstraction. The piece can be viewed as a whole, or worked in sections. Here I decided to focus on a work trip to LA, remembering the landscape, people and places by identifying colour, pattern and line. Words add to the story. It is presented with a wooden spool, gifted to me by Kaari from French General on my visit there.

Below: Anne Kelly, *Travel Scroll and Spool*. Mixed media textile

Right: Anne Kelly, *Travel Scroll and Spool*. Mixed media textile

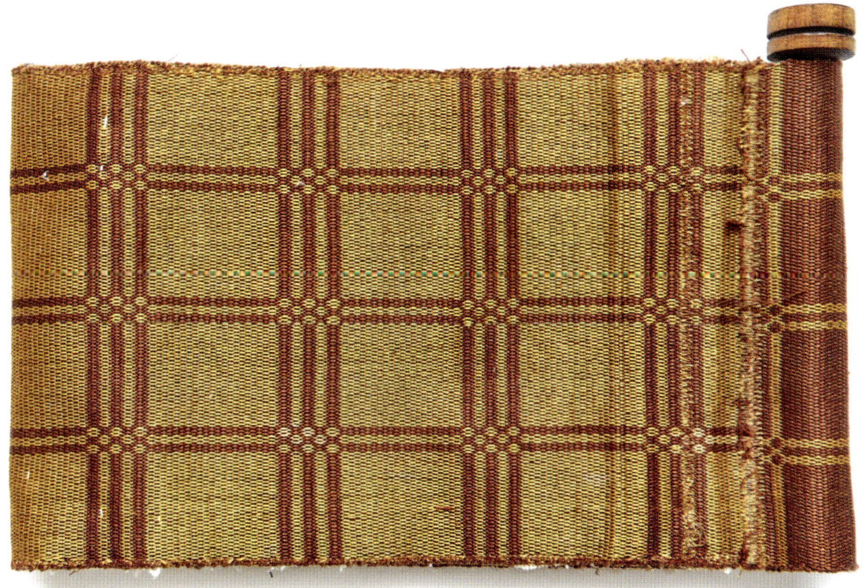

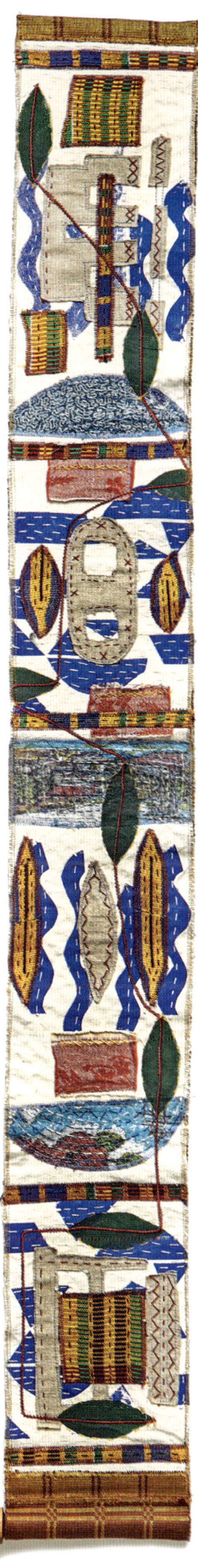

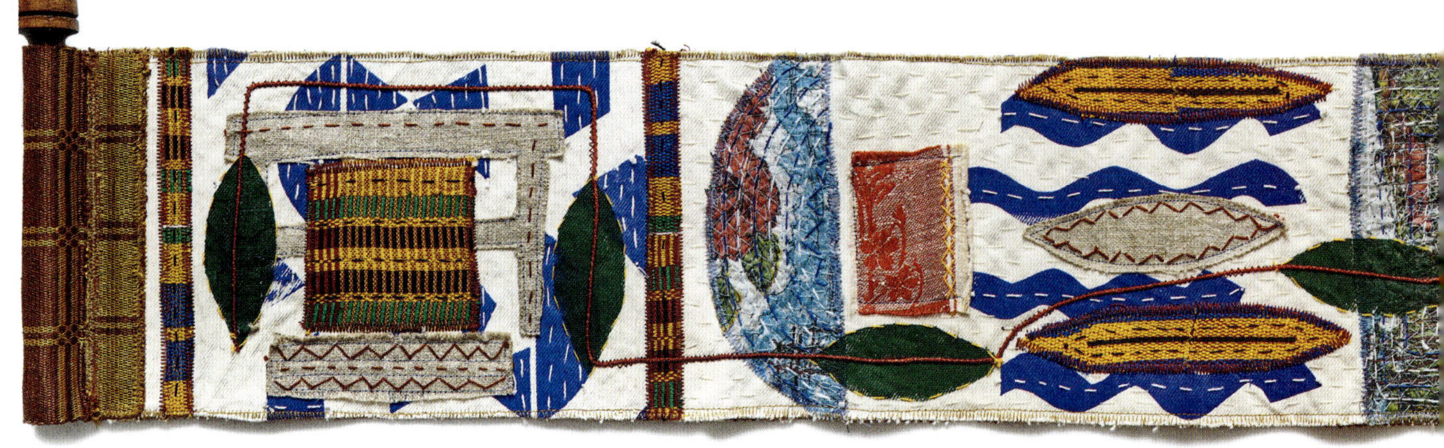

WORKSHOP

Making a travel scroll

Above & below: Anne Kelly, *Travel Scroll and Spool* (detail). Mixed media textile

MATERIALS REQUIRED

- A piece of aged or distressed cotton roughly 12 x 45cm (5 x 18in)
- Paint, dye or fabric colours to treat the surface
- Fabrics to create main shapes and objects
- Stitching kit, sewing machine if desired
- Small objects to add to piece and spool for presentation if you wish

METHOD

1. When approaching something as seemingly nebulous as this travel scroll, first compile a list or 'mind map' of thoughts that you would like to include on it.
2. Next decide on a colour or group of colours that work well with your theme. I chose bright blue to represent the Californian light of LA.
3. Colour your background, either as a solid colour or in sections, and leave to dry – I screen printed it (see page 65).
4. Arrange your pieces of fabric that represent shapes and impressions of your chosen trip on the surface, leaving space for stitch and writing. Take a photo.
5. Add any writing you wish at this stage.
6. Add stitching in three stages:
   a) architectural – to join pieces together.
   b) background – repetitive to add texture and interest.
   c) embellishment – to emphasise shape and add line.

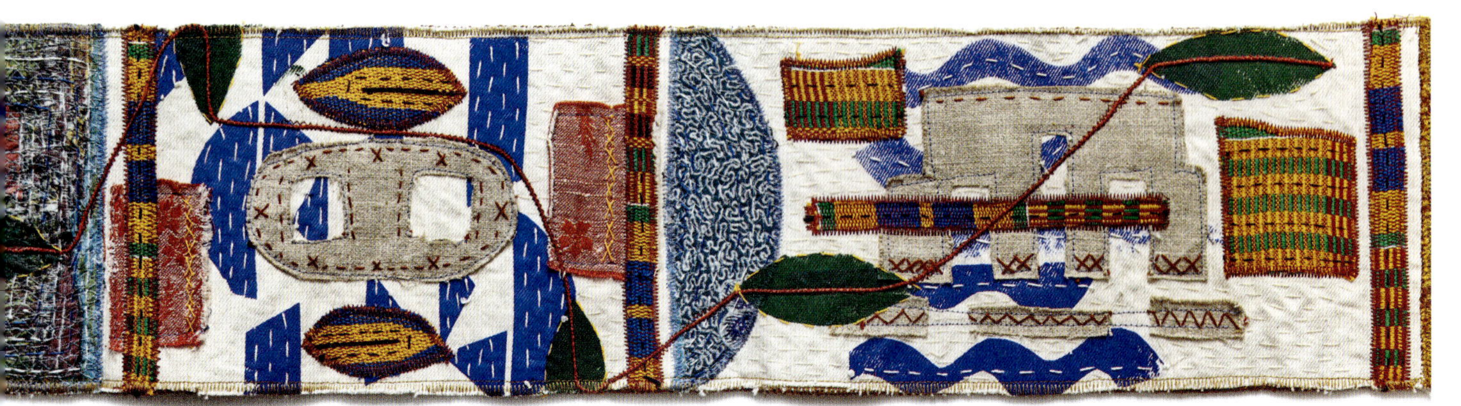

ENVIRONMENTAL

FEATURED ARTIST

Isabel Fletcher

Isabel Fletcher is an RCA graduate, working in London and exploring the waste in the fashion industry to create contemporary textile art pieces and site-specific installations. She describes her practice:

'I explore the way overlooked production offcuts can act as portals into the craft of their industry. Reframing disregarded offcuts as glimpses into the craft of making, I aim to increase empathy for possessions and encourage a reduction in consumption and waste. I see a direct connection between education, craft and sustainability. Whether interacting with my work at an exhibition, or attending one of my workshops, I hope to re-engage people with making and materials.

'I work intuitively by responding to the nuanced properties presented by offcuts, and my sculptural works take on ambiguous forms. I am fascinated by materials and their transformation from 2D to 3D. Combining offcuts with machine- and hand-stitch processes, I interact with the materials critically to identify their three-dimensional possibilities. The abstract nature of the work allows the mind to wander and imagine beyond the now normalised systems of take, make, waste. Imagination is key.

'*Industrial Offcuts: Freed of London* is led by the sourcing of production offcuts from the ballet and theatrical shoe manufacturer Freed of London. When visiting the workshop for the first time, I was struck by the contrast between the well-worn nature of the workshop and the gleaming beauty of the finished dance shoes. The maker of each component at Freed has perfected their craft, their bodies adapted and moulded to the rhythm of the repetitive processes. The offcuts themselves retain traces of this craft, each leftover scrap contributing to a silhouette of the making process.'

Below: Isabel Fletcher, *Industrial Offcuts* installation. Mixed media hanging

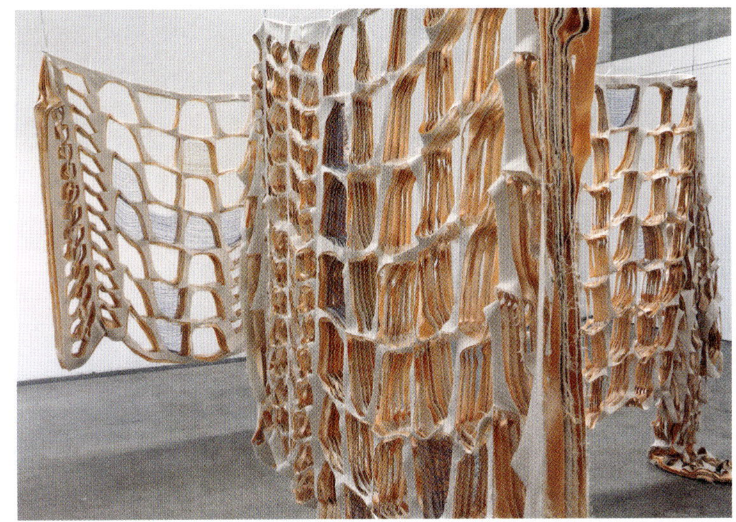

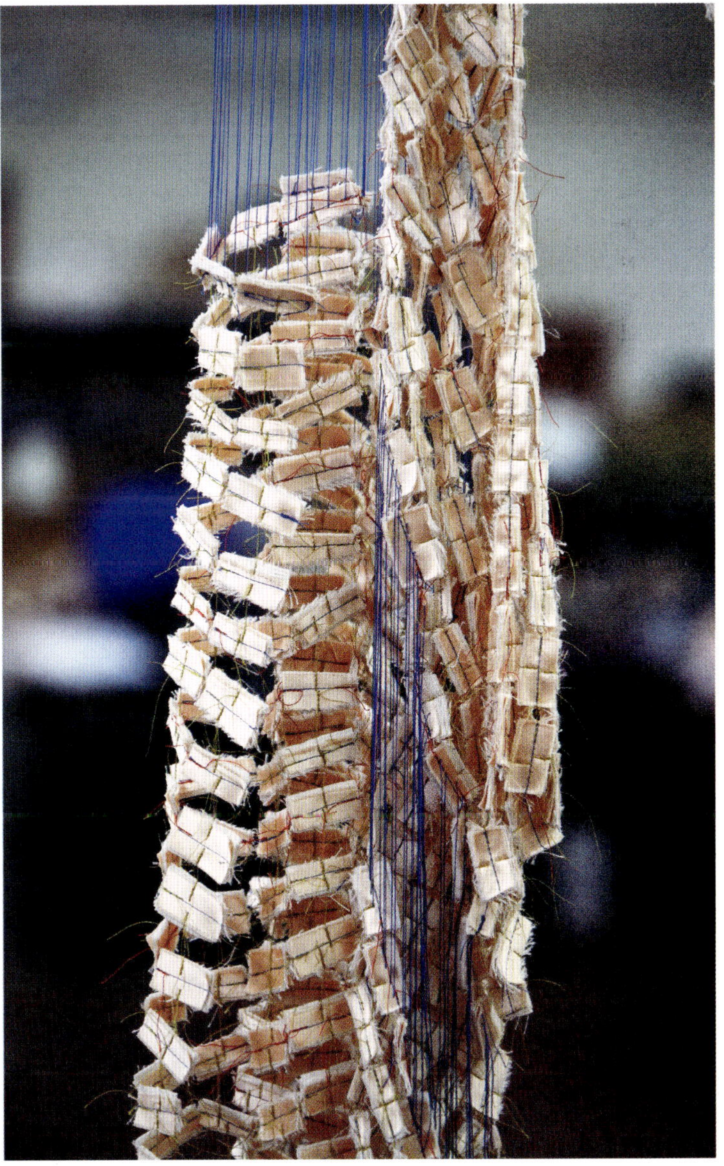

POLITICAL TEXTILES

In the seminal exhibition Unravel at the Barbican in 2024, we saw many examples and interpretations of work representing political ideals and expressions. Indeed, the exhibition itself was involved with a debate about the war in Gaza at the time. In the early twentieth century, the Russian Constructivists changed the course of modern art and textiles by refining it to its most minimal elements, focusing on shape, colour and line.

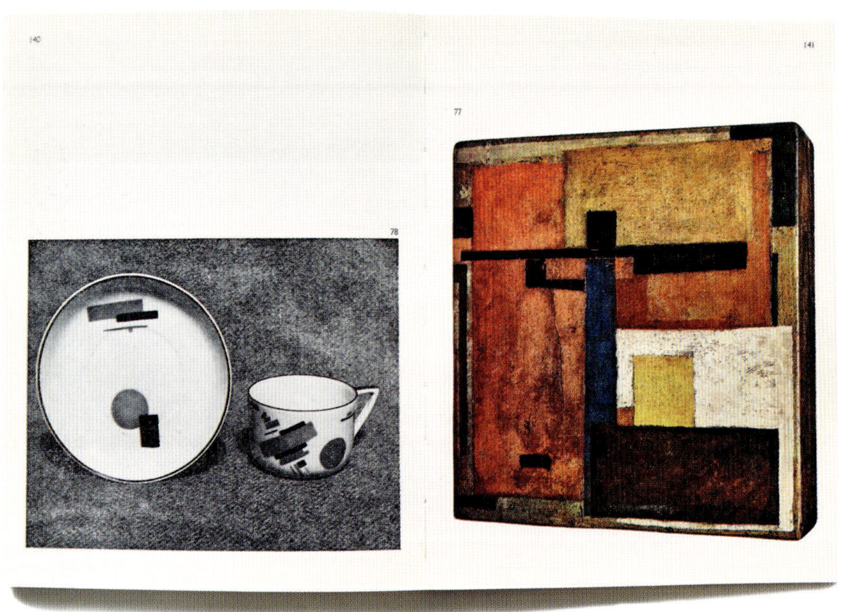

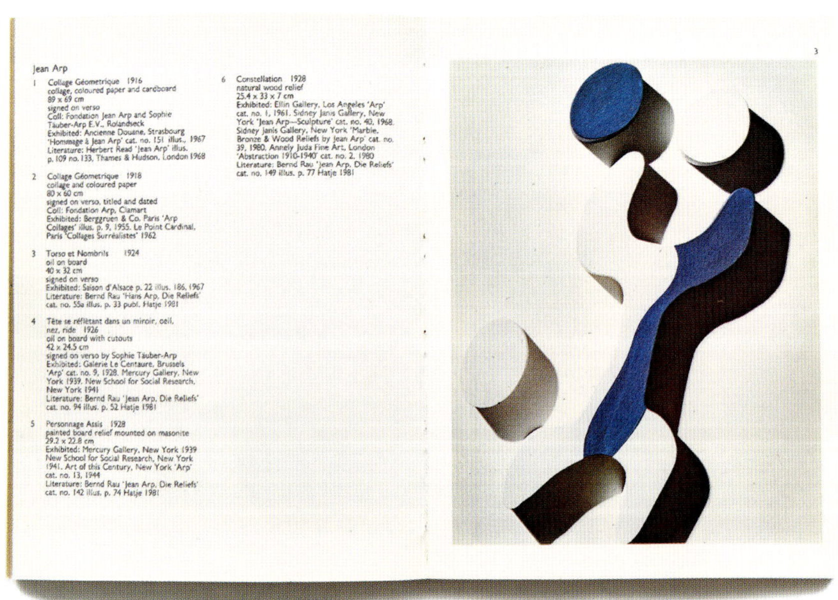

When I first came to the UK as a fine art graduate, one of my jobs was working with the amazing Annely Juda at her gallery in London. She introduced Constructivism to the UK art market, and her son David still runs the gallery very successfully. I saw at close hand the beauty of these originals and helped to compile the gallery catalogues for exhibitions as well as carrying out sundry tasks. The gallery has very kindly given permission for the reproduction of this piece by Alan Reynolds, seen opposite.

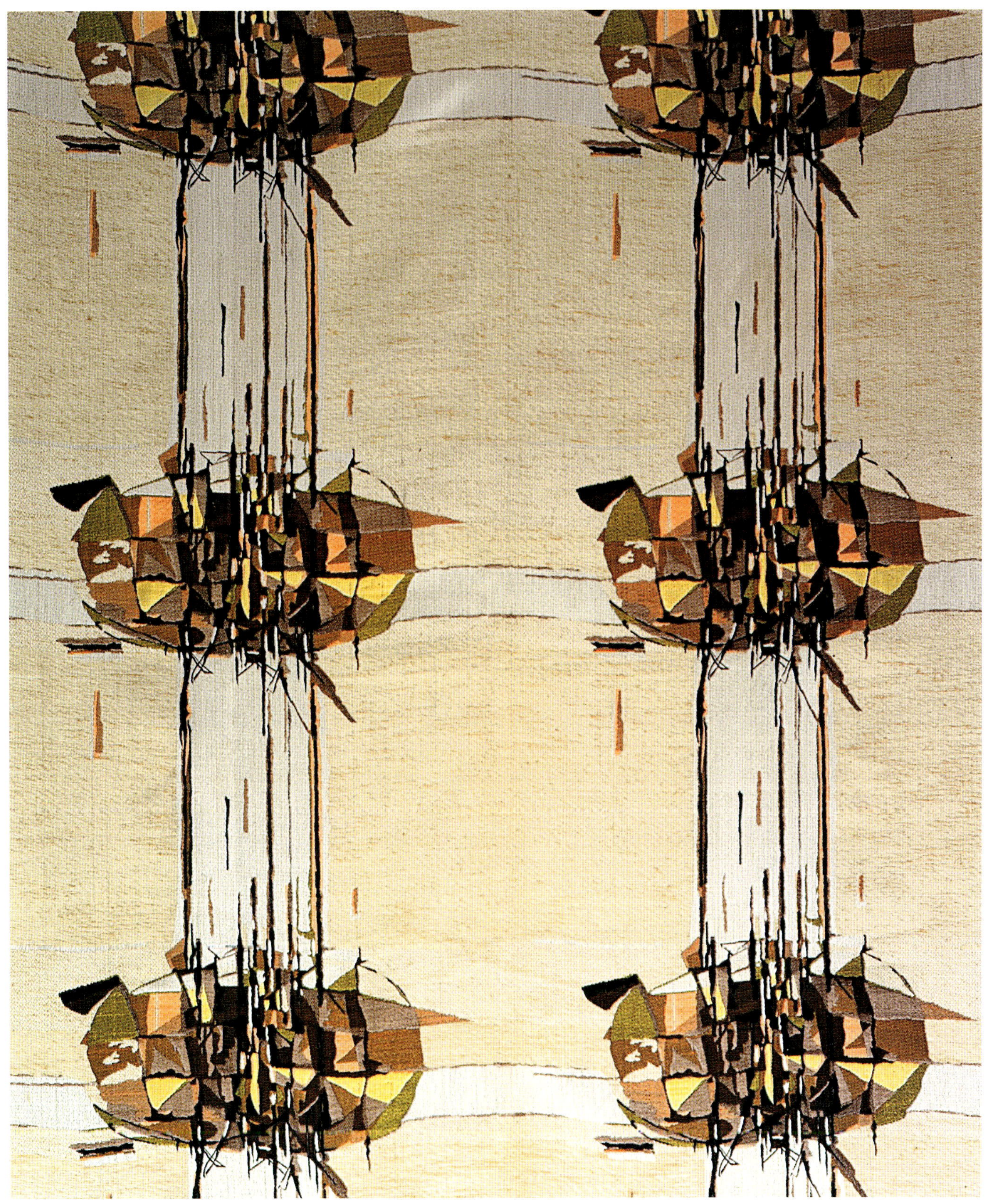

Opposite: Annely Juda Fine Art, catalogue pages

Above: Alan Reynolds, *Crystalline Image* (for Edinburgh Weavers Ltd). Jacquard and woven cotton and rayon

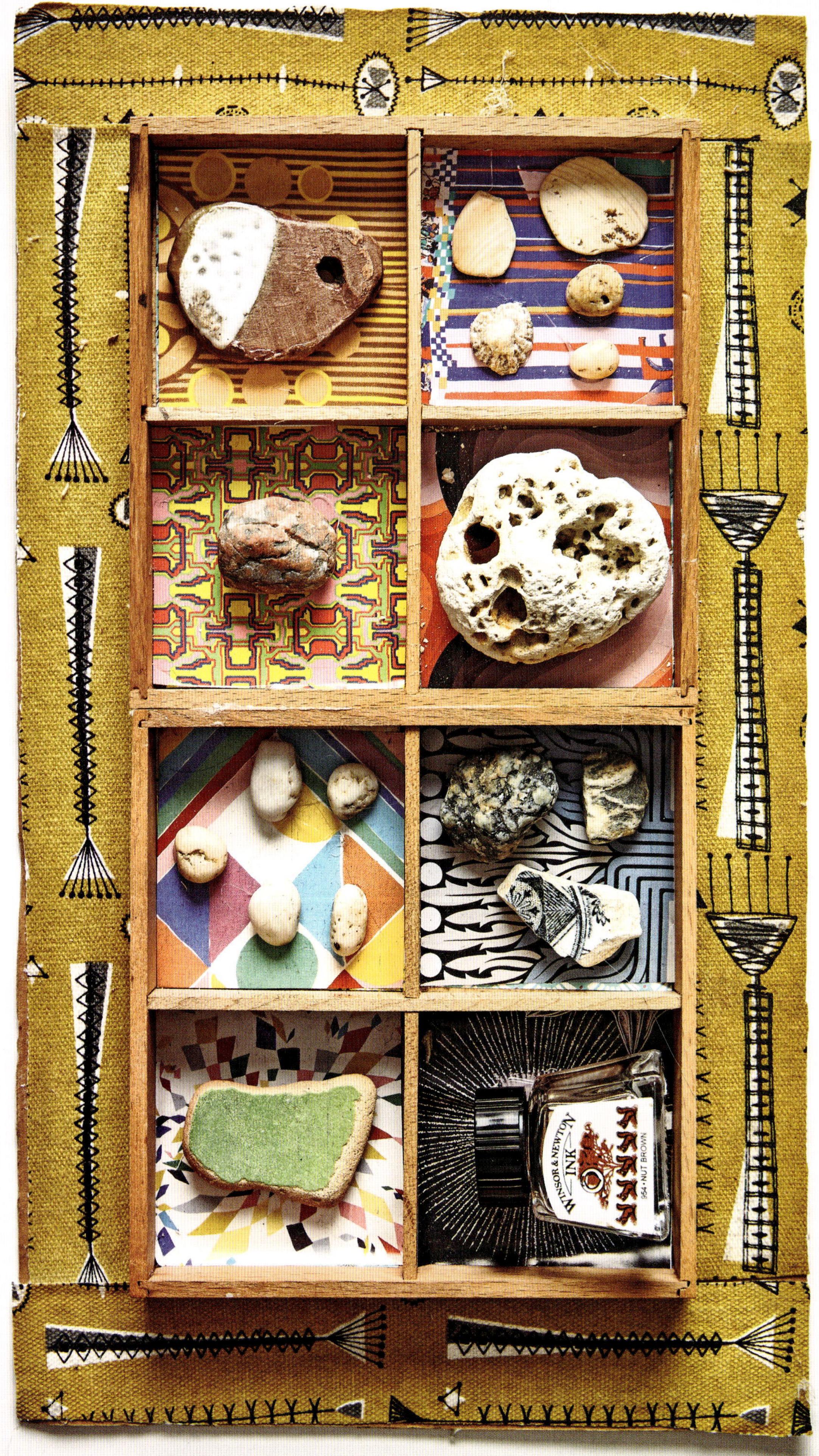
WINSOR & NEWTON
INK
954 • NUT BROWN

Chapter 6

PLAYING AND PROCESS

'The point isn't to know what you're doing. The point is to have an experience doing something.'

Kiki Smith (b.1954)

Opposite: Anne Kelly, *Inspiration Box, Sea Studio*. Mixed media

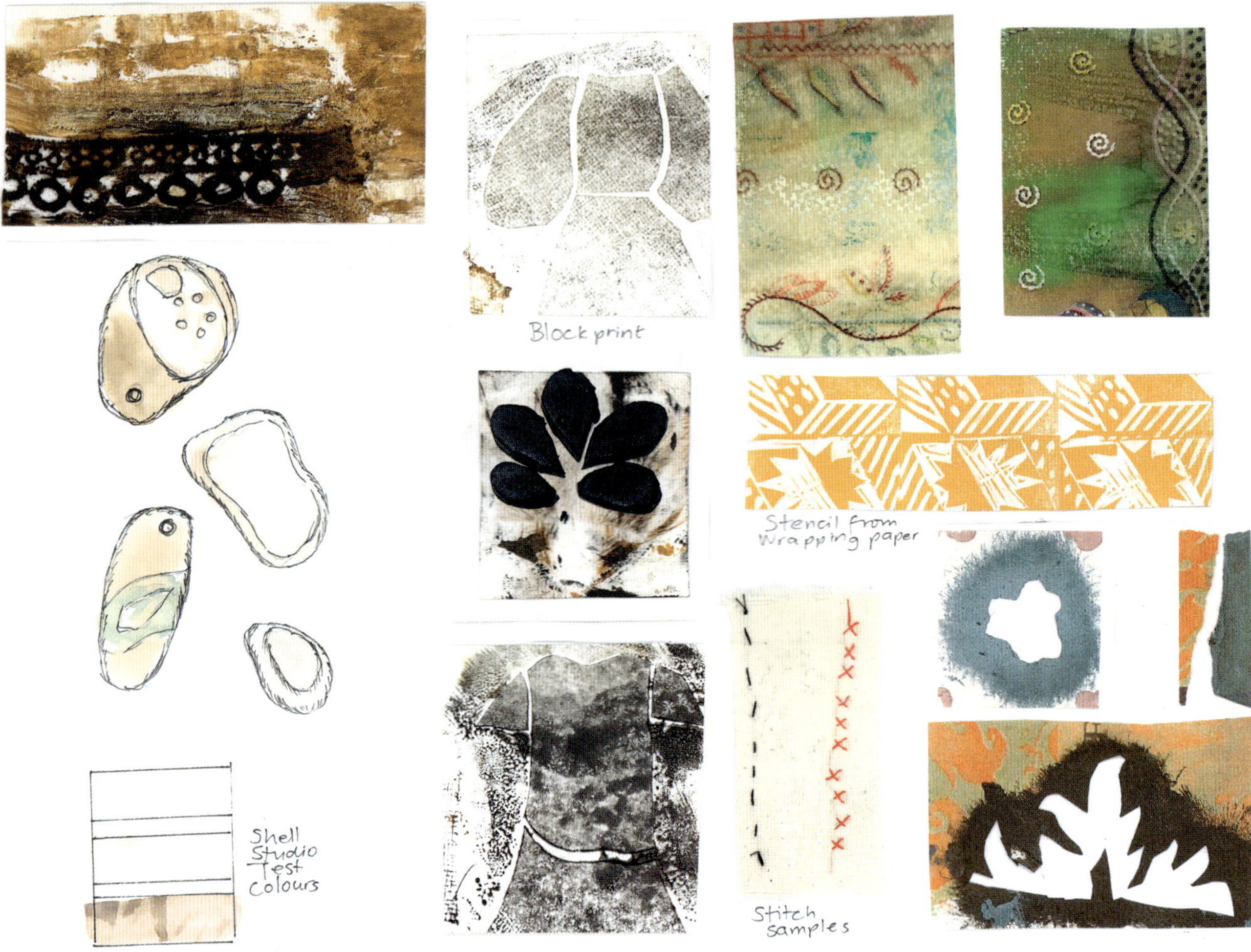

CREATIVE PLAY

In this chapter, we explore process and making, looking at the studios and practices of artists who enjoy the challenges and experimental nature of creating original abstract work. The boundary between paper, cloth and even sculpture has been deliberately blurred to enable you to consider as many options as possible.

It is wonderful to have a dedicated space to work in, but as someone who only got a studio space relatively late in my practice, I appreciate that not everyone does. You can carry out a huge variety of processes at a table anywhere in your living space. If you start with sampling and smaller work, you can scale up when more confident.

My practice is part of my life as a whole and has always had to fit around a busy family. I appreciate that this is the case for many, and along with caring duties, not everyone has the time mentally or physically to start new work. I have intentionally simplified the 'how to's' in this book for this reason.

Above: Anne Kelly, *Sketchbook Pages*. Mixed media

Opposite top: Materials for making ink – tea bags, spoon, mixing cup

Opposite bottom: Anne Kelly, *Kitchen Ink*. Handmade ink on paper

'KITCHEN' INK

I was inspired to make homemade ink, using kitchen ingredients, after seeing lots of experiments by a range of practitioners. I thought I would try the easiest and most accessible method – tea bags – to create a homemade ink.

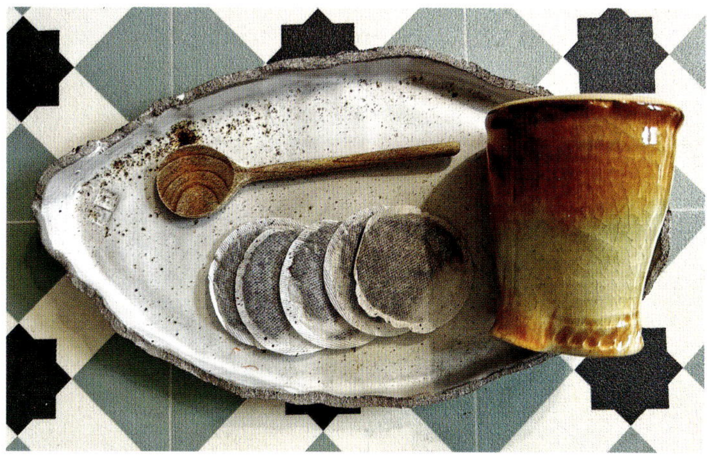

WORKSHOP

Making kitchen ink

MATERIALS REQUIRED

- Four or five black tea bags
- Two tablespoons honey
- Boiling water
- Clean, empty jar
- Coffee filter and one cup filter holder (if desired)
- Brushes and paper for testing

METHOD

1. Boil the kettle and make a very strong mug of tea with four or five teabags. Set aside.
2. When the tea is cool, remove the bags and add the honey, then stir well. The stronger the tea, the darker your 'ink' will be.
3. Strain the tea mixture through a coffee filter paper if desired to remove any 'bits' that may be there.
4. You can now test your 'ink', using a brush or sponge. My samples are to the right.

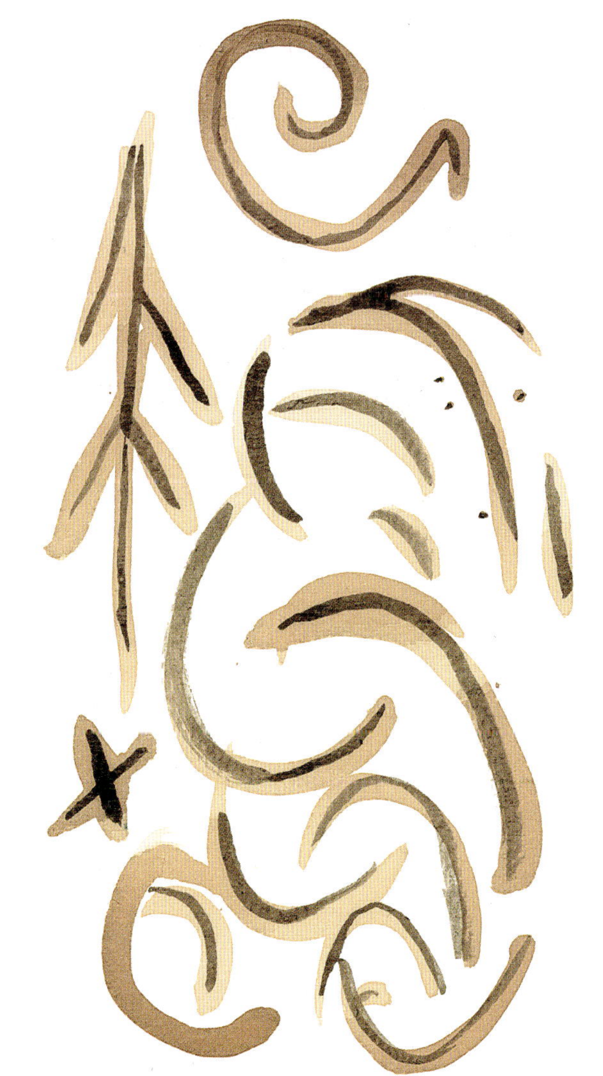

FEATURED ARTIST

Lorna Crane

I met Lorna Crane on my first visit to Australia in 2017. She was most welcoming when we dropped into her studio to see her work and brush-making. Lorna says:

'My abstract landscape-inspired works, in many shapes and forms, speak about moments of time where narratives are formed, via my own visual lexicon in an experiential and gestural manner. It is from that place from deep within where shapes form and are distilled into fragments of past and present, merging together in an abstract form. It is about seeking questions and revealing an intimate personalised glimpse into my inner landscape – both physically and metaphorically – known and unknown, from the land and of the land.

'I moved to my new studio last year and like to use inventive painting techniques that combine collage, patina effects and surface pattern. I became known as "the Brush-maker", sourcing natural fibres and materials from my local environment. I have been teaching my unique way of exploring visual language through live workshops all around Australia and most recently online internationally.'

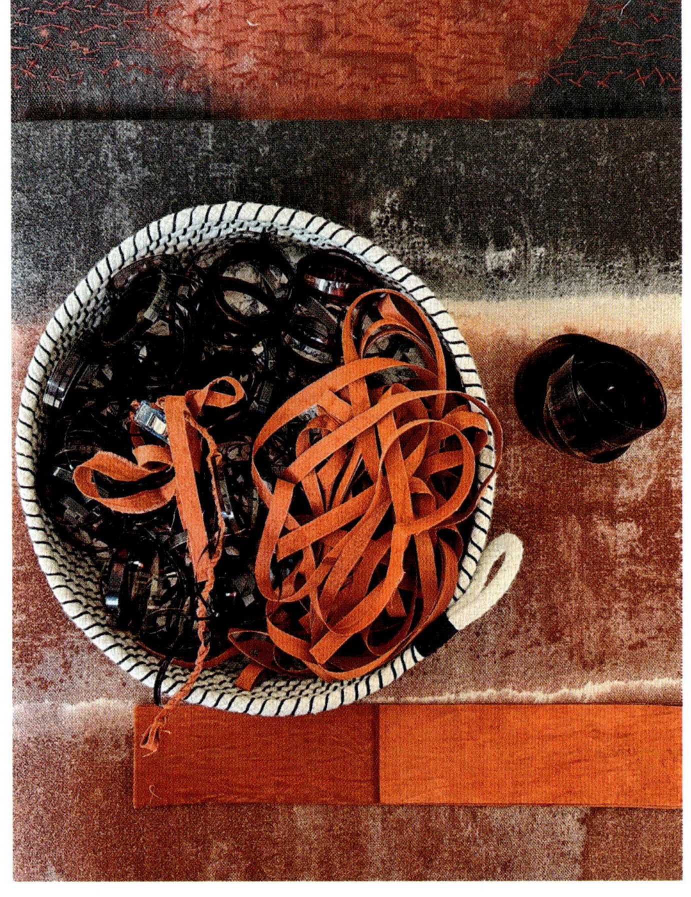

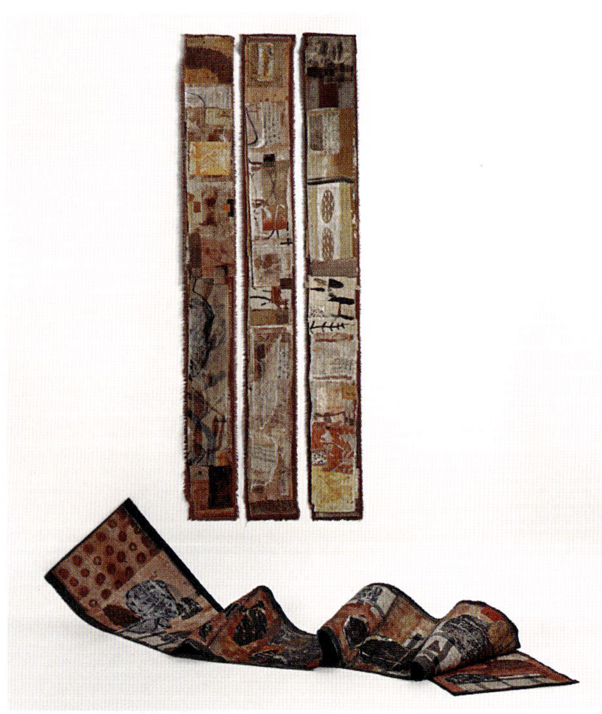

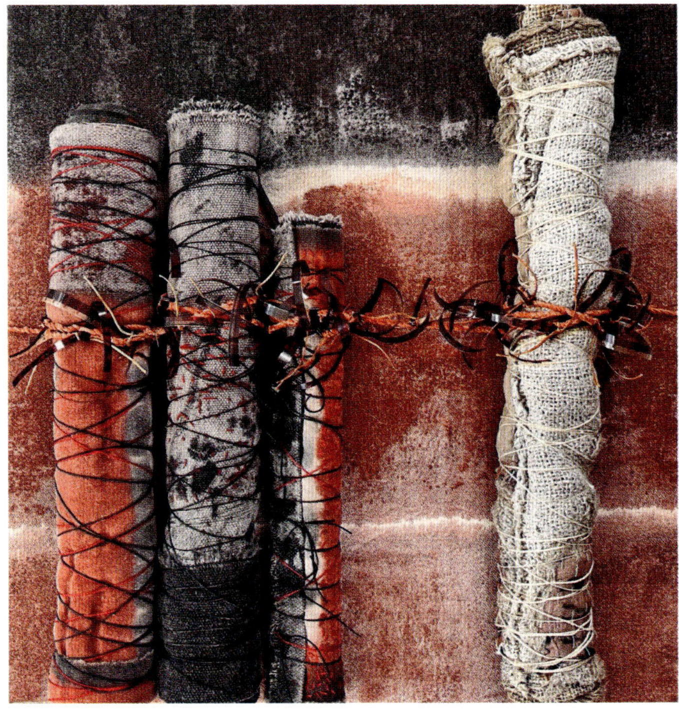

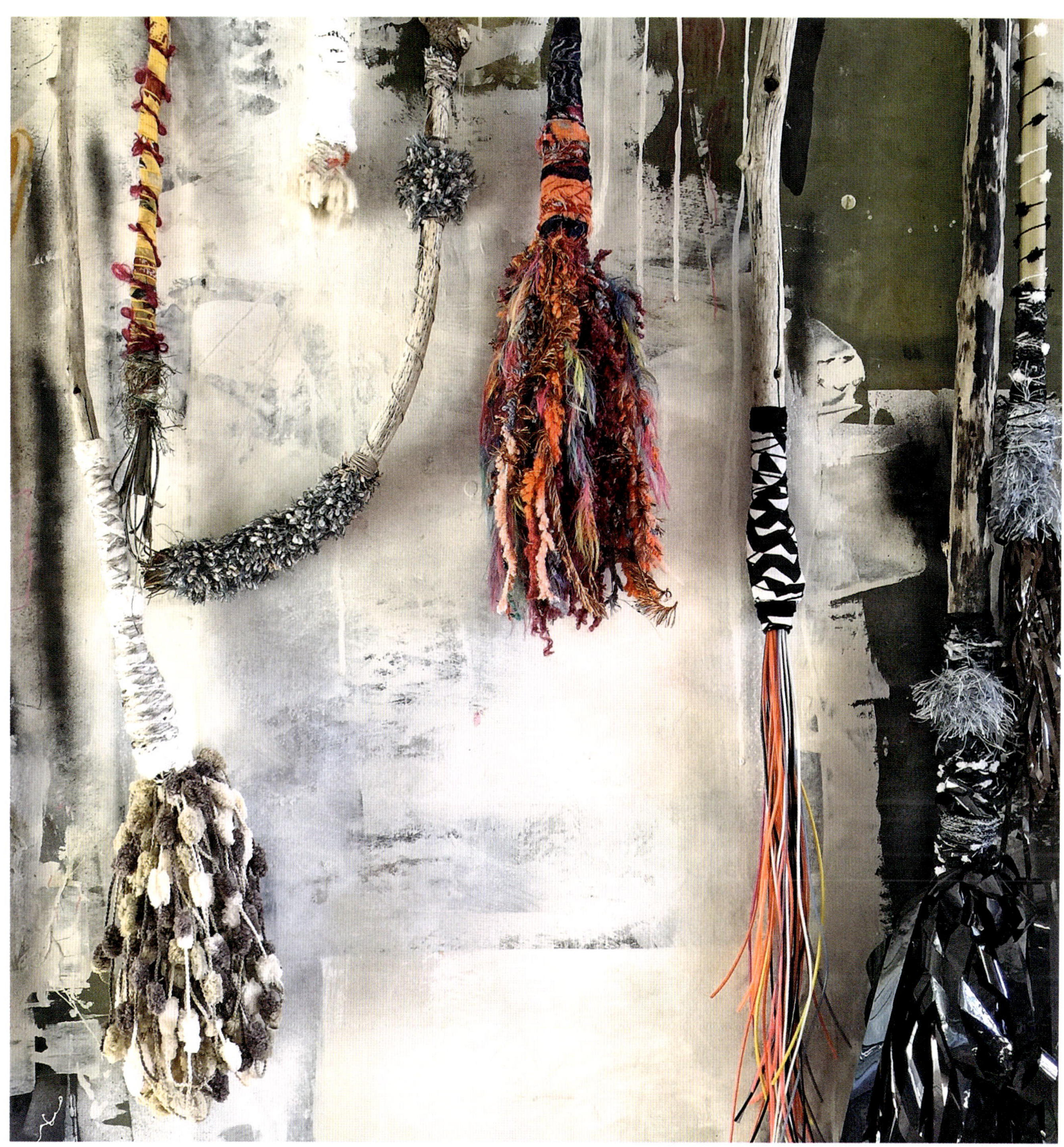

Opposite left: Lorna Crane, *Symbiosis Series*. Mixed media installation

Opposite right (above & below): Lorna Crane, textures for *Reprise*. Mixed media textiles

Above: Lorna Crane, handmade brushes

NATURAL MATERIALS FOR BRUSHES AND MARK-MAKING

Your garden or local park can provide a wealth of natural materials to make rough brushes and other tools for mark-making. I will show you how to make a simple brush, using found materials.

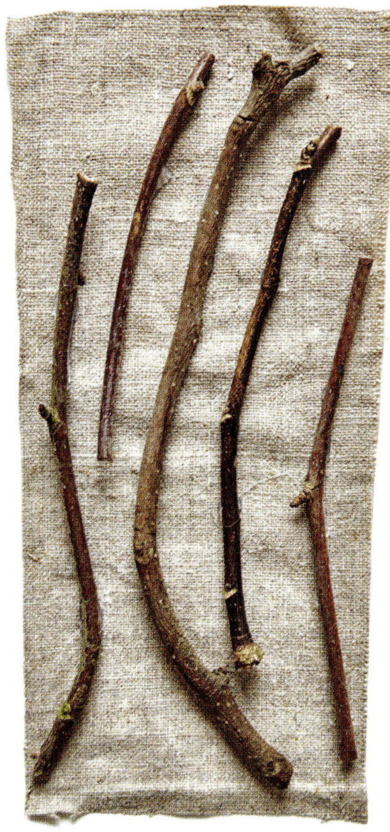

WORKSHOP

Making a simple brush

Right: Anne Kelly, materials for brush-making. Twigs and willow

Below left: Anne Kelly, handmade brush. Twigs, willow and calico

Below right: Anne Kelly, *Handmade Brush Ink Test*. Homemade ink on paper

MATERIALS REQUIRED

- Softer twigs, plant stems, willow branches (all dried). Look for a range of lengths and textures – around twenty minimum. Each should be at least 10cm (4in) long
- A larger stubby piece of wood (found or tree branch, cut) for a 'handle'
- Scissors or secateurs
- Twine and rubber band or clamp
- Torn cotton rags in thin strips, PVA glue (optional)

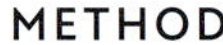

METHOD

1. Assemble your softer twigs and plant materials and select the best and strongest pieces. You will need around twelve to fifteen depending on how big you want your 'brush' to be.
2. Tie your pieces with twine, using a rubber band or clamp to hold them together. They need to be tied tightly with a secure knot.
3. Now you need to attach them to your 'handle'. Wrap twine around the twigs and join them securely onto the handle. You will need to make several layers of wrapping. You can also use strips of rags and a bit of PVA glue if the string isn't strong enough, or to add extra stability.
4. Now you can test your 'brush', with paint or ink, to create interesting lines and shapes.

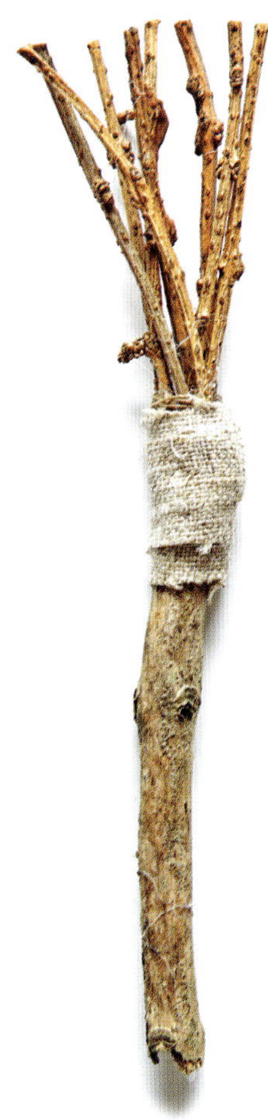

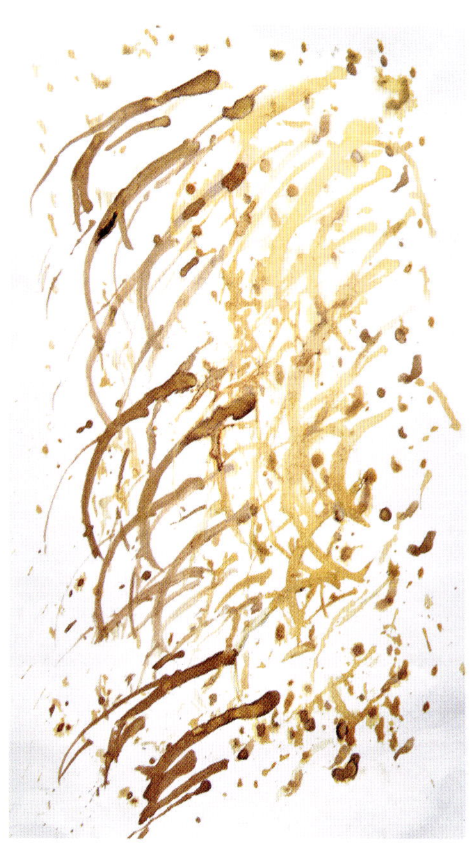

Here are some natural materials that you can use to create with:

- Plant fibres (both fresh and dried materials)
- Flower heads
- Leaves
- Branches
- Twigs
- Feathers
- Hand spun wool
- Knotted yarn
- Ocean debris

Lorna Crane also makes beautiful brushes in her studio, as you can see on page 109.

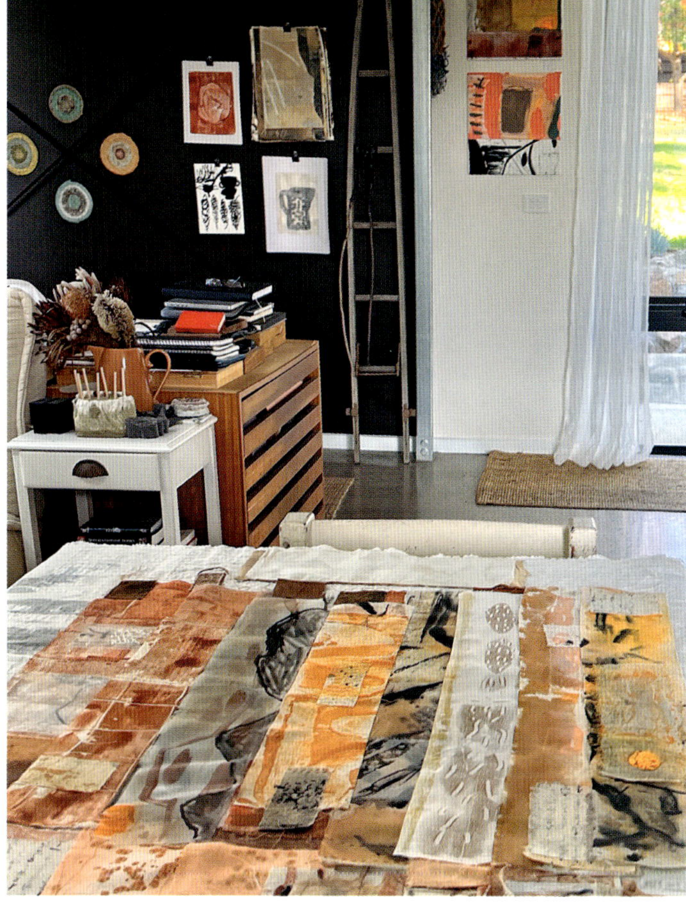

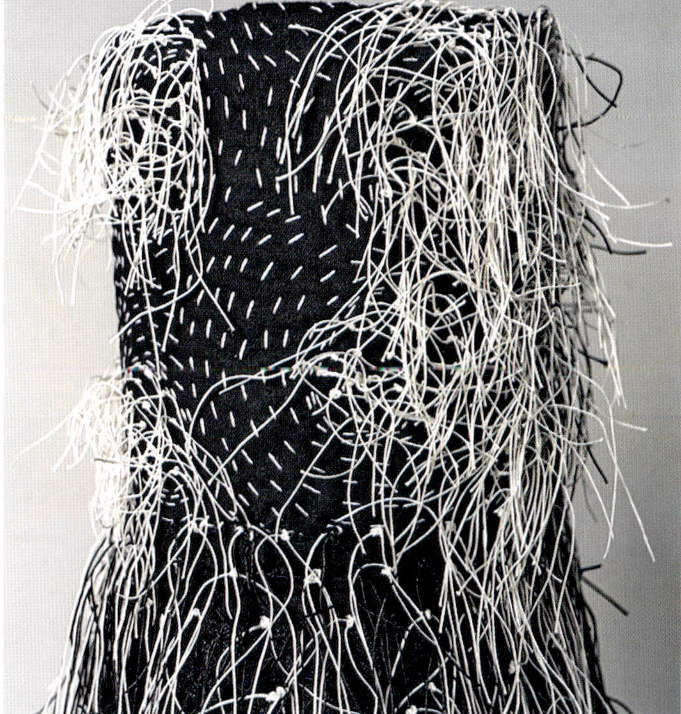

Above right: Lorna Crane, studio shot

Right: Lorna Crane, *Infinite Longing* (vessel detail). Mixed media

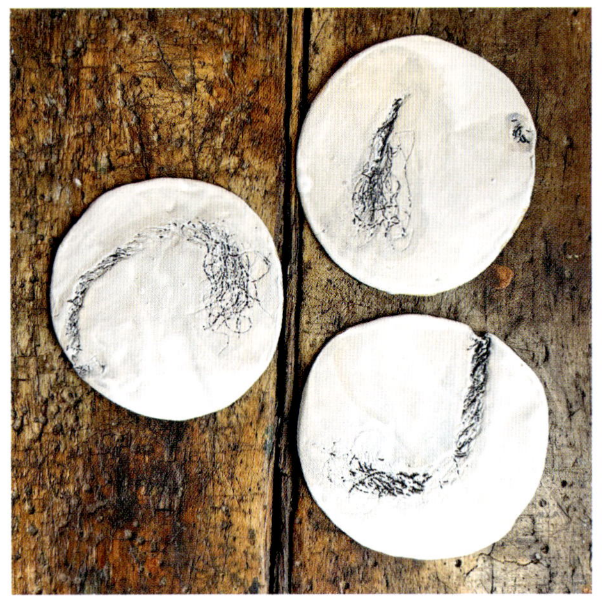

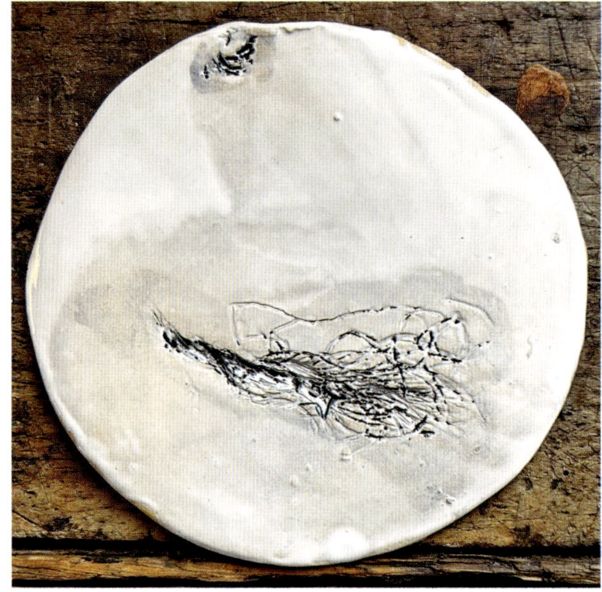

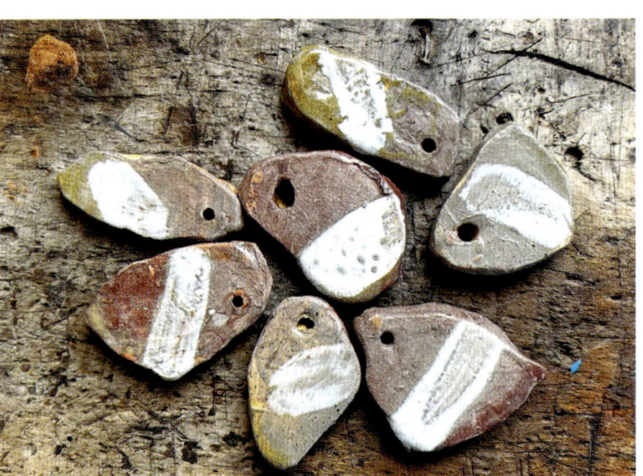

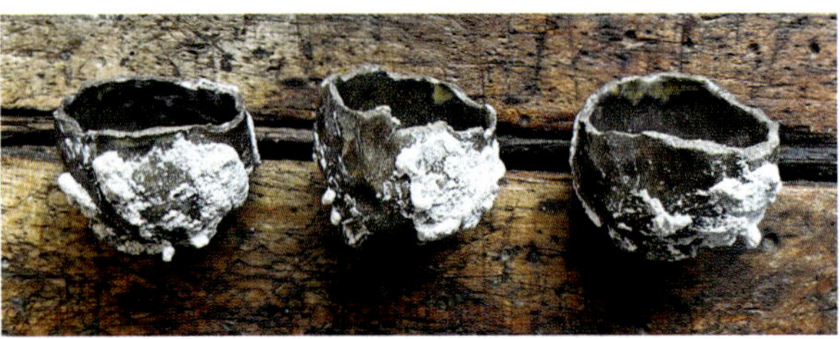

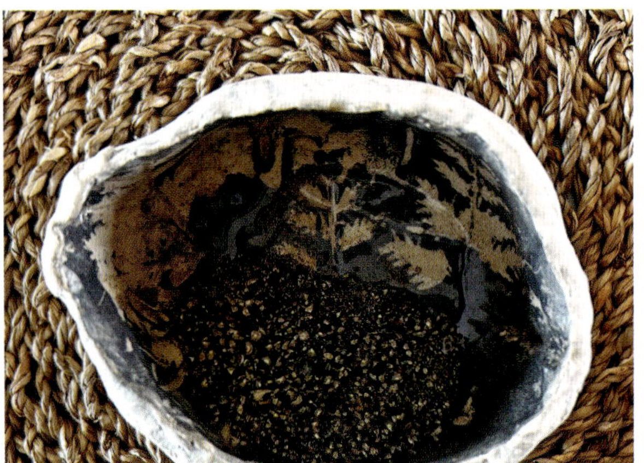

From top: Heather Gratton, *Pressed Plates* and *Pressed Plates* (detail). Printed ceramic

Shell-inspired Tags. Glazed ceramic

Untitled Vessels. Ceramic

Untitled Vessel. Ceramic

SEA STUDIO

FEATURED ARTIST
Heather Gratton

Heather Gratton is a fashion-trained interior designer and journalist who also makes beautiful ceramics for the home. She lives beside the sea in Shoreham, West Sussex. Her house is a showcase for her work, which reflects the colours, textures and light of the sea. Her work is inspiring and the patterns in it translate well into abstract textile art experiments.

WORKSHOP

Making an abstract sea-inspired collage

MATERIALS REQUIRED

- A selection of sea-inspired natural forms; stones, shells, seaweed
- Sheet of cartridge paper and pencil, paper scissors, Frixion pen (if desired)
- A square or rectangular piece of heavy cotton or linen, 21 x 28cm (8 x 11in)
- Watercolour paint, ink and/or dye, brushes
- Small pieces of fabric for appliqué in tonal colours
- Glue stick
- Hand-stitching kit, sewing scissors
- Masking tape, paper or plastic to protect surfaces

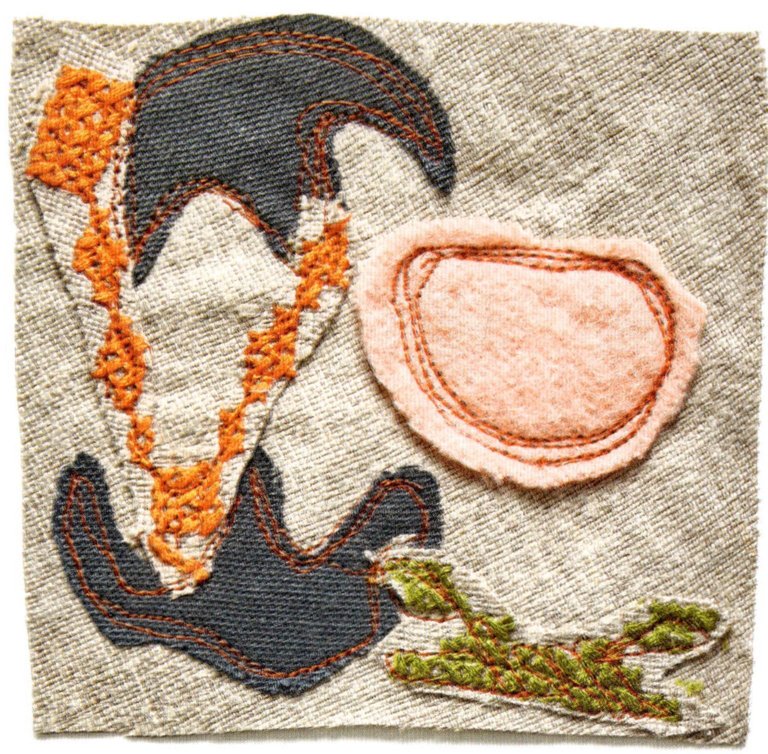

METHOD

1 Draw the sea-inspired shapes onto your cartridge paper, overlapping and covering most of it with pencil outlines.

2 Tape your fabric to a flat smooth surface with paper or plastic underneath to protect your table.

3 Cut out your shapes from the design and trace around them onto your fabric.

4 Paint the shapes in a variety of shades using a watery ink, paint and/or dye. Use a dry brush so the paint doesn't spread too much. Leave to dry.

5 Cut small pieces of fabric to place on your design. I like to highlight lighter areas and markings. Glue them onto the surface.

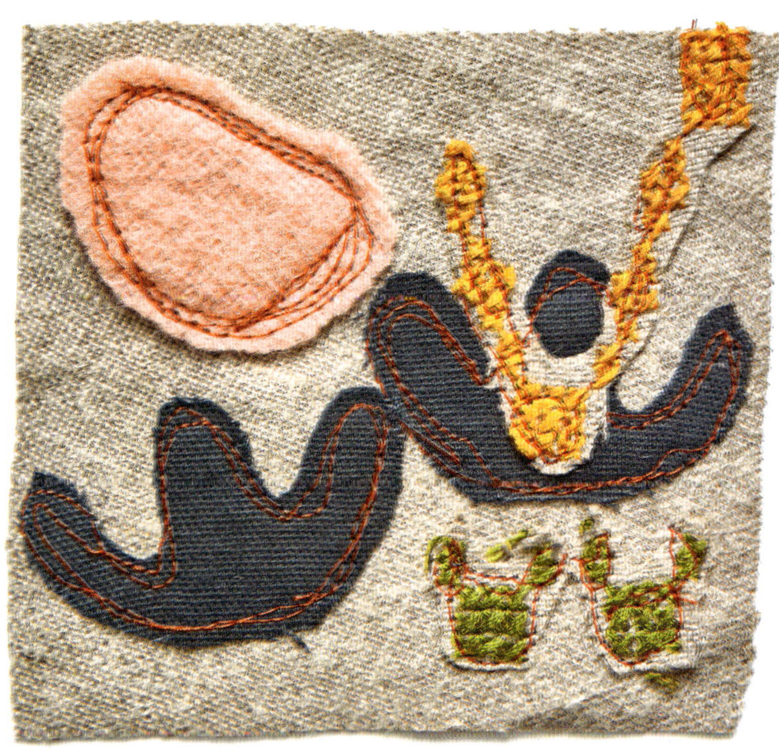

Above: Anne Kelly, *Sea Collages*. Mixed media textiles

FEATURED ARTIST
Deborah White

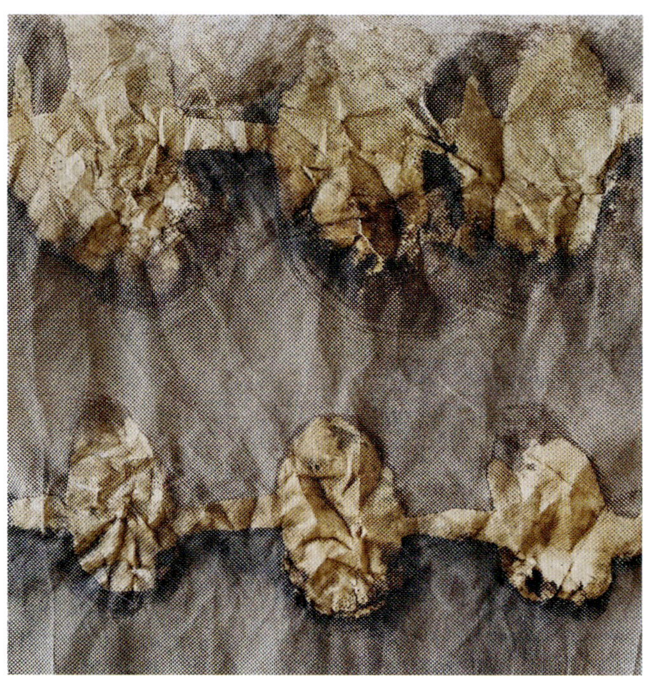

Above: Deborah White, *Untitled*. Mixed media textile

I met Deborah White through her filming work. She is a curious and thoughtful artist and maker, eager to explore materials and techniques, as she says here:

'I am an artist committed to pushing the boundaries of cloth through the exploration of recycling and layering unconventional elements in fearless experimentation. Studio play, for me, is where wild explorations are redefined, discarded materials find new life, and experimentation knows no bounds. At the heart of it all, I am, first and foremost, a textile artist, weaving stories and emotions into every layer, whether it be 3D, 2D, or large flat work. But my canvas extends beyond fabrics, encompassing a diverse array of materials sourced from the remnants of our collective consumption.

'I'm drawn to layering with unconventional building materials. This choice is fuelled by the urgency I feel amid the rapid construction pace in our ever-expanding urban landscapes. As builders move swiftly from project to project, leftover remnants accumulate at an alarming rate. These discarded resources become my palette, infused with the untold stories of their previous owners. As I incorporate these elements into my work, I not only breathe new life into forgotten remnants but also confront the viewer with the stark reality of our throwaway culture.

'Amid this eclectic blend, recycled fabrics serve as the beating heart of my practice. Each piece carries with it a history, a tapestry of memories. From discarded garments to forgotten remnants, these fabrics speak of journeys taken and stories untold, waiting to be reshaped into narratives of beauty and resilience. My artistic process lies between intuition and intention, guided by an insatiable curiosity to push the uncharted territories of possibility. Central to my practice is the belief that art has the power to spark dialogue and provoke introspection.'

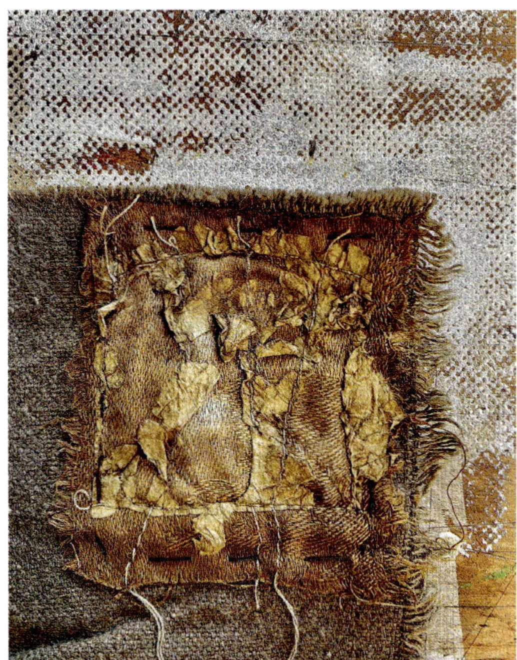

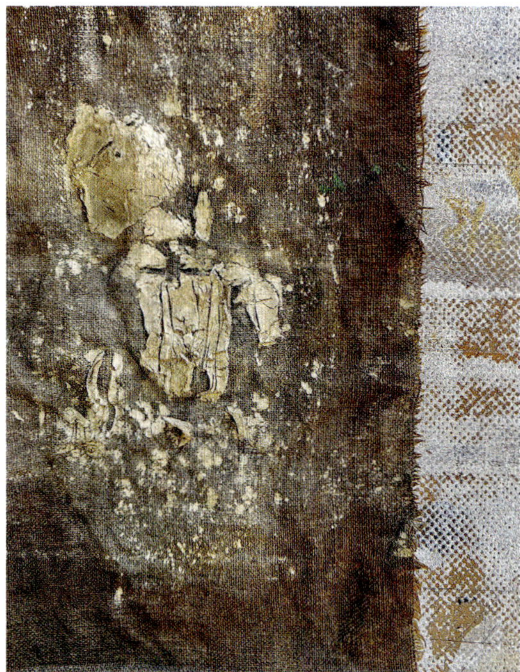

Right: Deborah White, *Untitled*. Mixed media textile

Far right: Deborah White, *Untitled*. Mixed media textile

PAPER COLLAGE – FROM PAPER TO CLOTH

It is useful to make work in paper and to explore shapes and designs before committing to cloth. They can also become artworks in their own right. Hunt through your collections for different types and weights of paper, card and tissue.

Materials you can use:

- **For background** – old book covers, card boxes, canvas, postcards
- **For layering** – book pages, magazines, flyers, tissue paper, foil sweet wrappers
- **Adhesives** – glue stick, PVA, rubber cement, paper paste
- **Tools** – paper scissors, pencil, cutting board and scalpel, rubber stamps and ink, hole punch, ruler

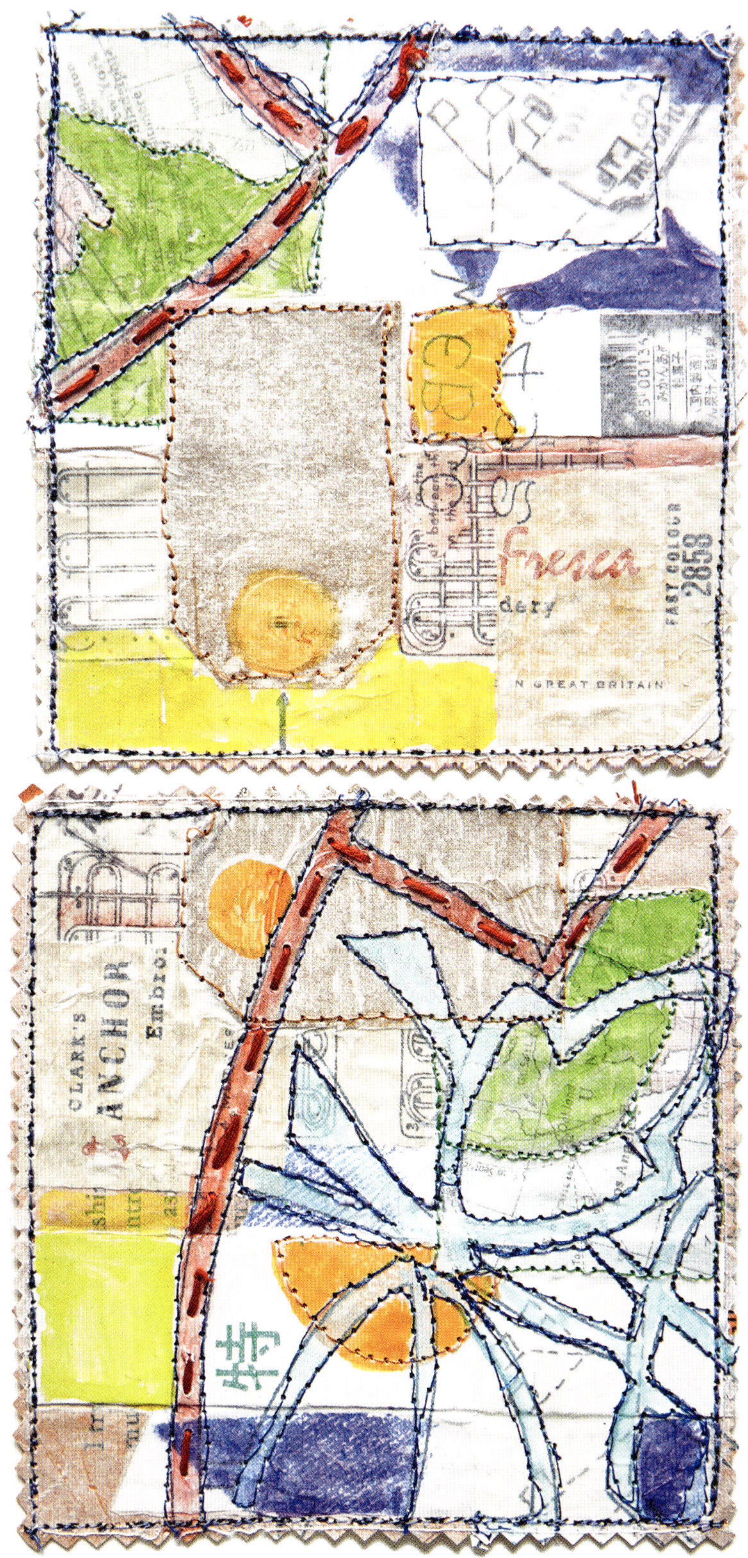

Right: Anne Kelly, *Paper Collage*. Mixed media

FEATURED ARTIST

Sarah Z. Short

Sarah's work is minimal and powerful and she works in series, making new meaning from discarded ephemera. She says:

'I'm an artist and printmaker from Rhode Island, where my work is inspired by the calm of the ocean and the woods. Old papers and books are my inspiration and my medium. You'll find me at estate sales, digging into boxes in the basement and the bottom drawers of desks finding the papers that most people would discard. My work seeks to create a new narrative for found vintage ephemera and papers. Instead of being overlooked or discarded, I rework these papers into contemporary art so that when someone looks at my collages, they will recognise the materials, explore the layers, and find the bits of text that I've included or abstracted. Maybe this will spark a memory or cause them to rethink the value of the old books and papers we casually discard.

'I have a vintage letterpress and a large collection of wooden type and will spend a solid day or two printing onto vintage paper. My collages begin when a certain paper sparks my interest. I have flat files, bins and drawers full of sorted papers. I might choose a file folder full of those letterpress papers as my starting place. I work abstractly, cutting and tearing as I go and incorporating additional fabric and materials as the composition develops. My collages are layered as I soften edges, cover text, and create a piece that gets more interesting the closer you look. I don't use many complex shapes because I like torn edges and blocky shapes. I incorporate fabric from book covers and maps and the threads are left uncut, creating texture and line.'

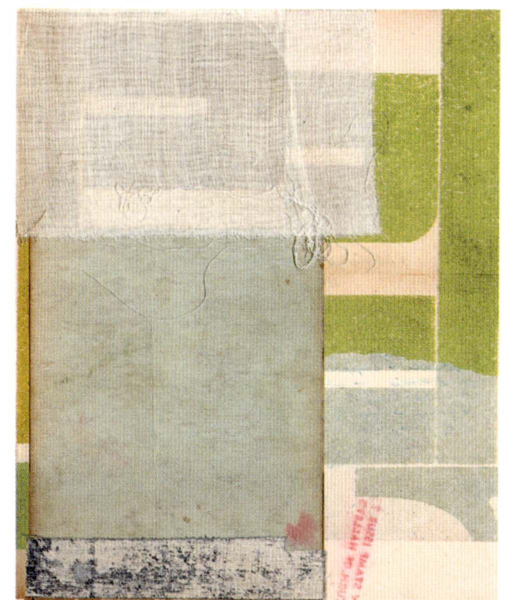

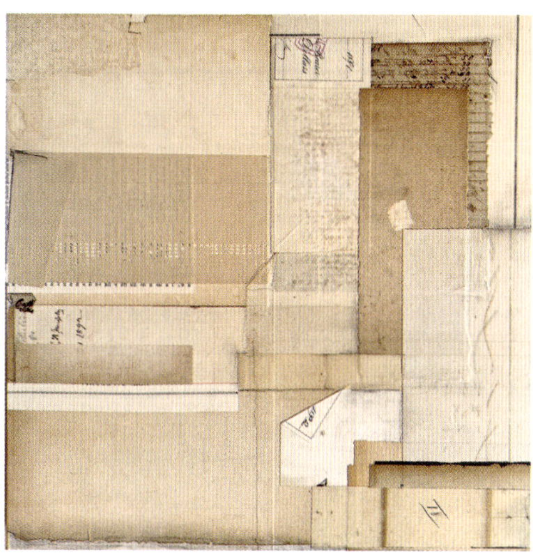

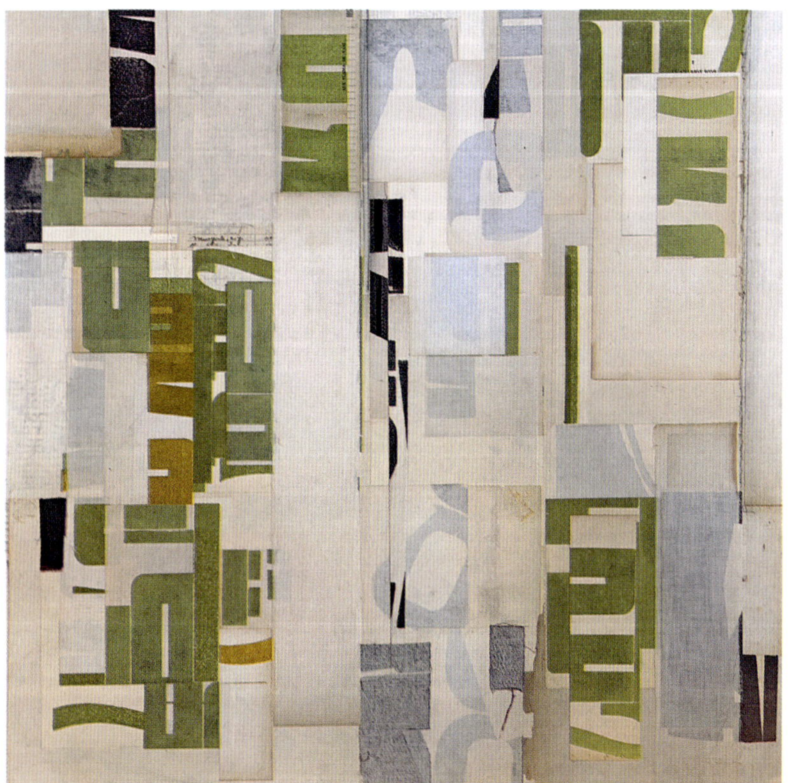

Top: Sarah Z. Short, *Halley's Comet*. Mixed media collage

Above: Sarah Z. Short, *Genius*. Mixed media collage

Left: Sarah Z. Short, *Beneath*. Mixed media collage

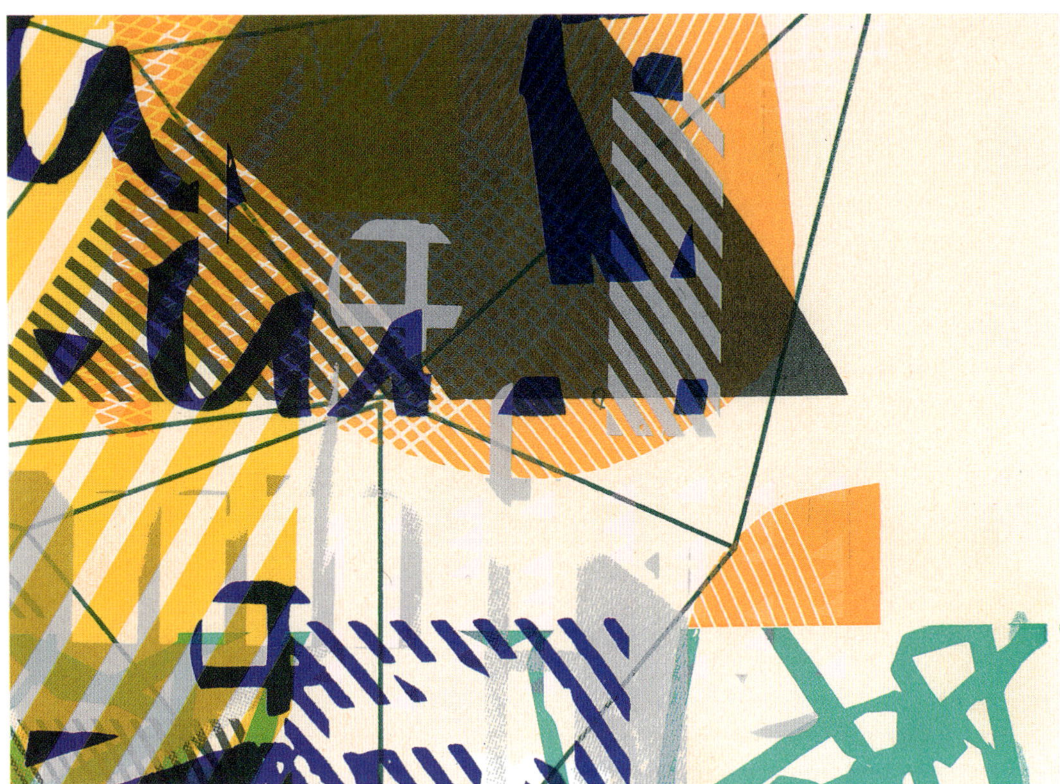

Left: Iain Perry, *Screen-printed Card*. Screen print

Below: Iain Perry, *Dope Moire Mirrorball*. Screen print

SCREEN PRINTING ON PAPER

I have given instructions for screen printing in Chapter Three (see page 64). For printing on paper, you can use the same printing ink or buy paper printing ink, which comes in a large array of colours. A good, flat, slightly heavier cartridge paper is ideal for printing on, but many artists use wood and alternative surfaces for experimentation. As you gain experience you can use a variety of techniques.

FEATURED ARTIST

Iain Perry

Iain Perry creates deeply layered, colour-saturated, abstract screen-printed collages. His fearless use of colour (and the language he uses to describe it) is something we can all appreciate. Here he explains his process:

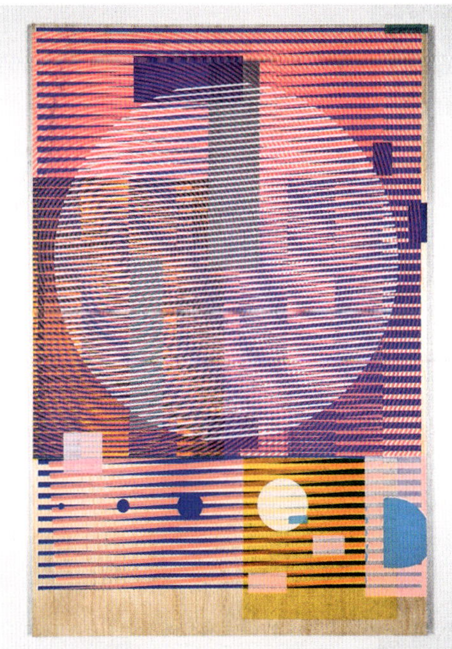

'The image generation is a slow process of experimentation, of trial and error; collecting imagery, developing patterns and overlaying all the different elements until new and intriguing relationships emerge. The results are brightly lit beacons of Balearic Zen shining out amid a visual, digital landscape that is in constant upheaval, upgraded and updated daily.

'My work borrows language from other art forms – film and photography, poetry, electronic music and rave culture – amalgamating, mixing and remixing, offering up familiar signs and signifiers but now refracted and reflected, illusive and allusive, all lightly seasoned with a slightly understated degree of subversion. Layers of fragmented iconography and geometry coalesce on the paper, like motes of dust trapped in a beam of light teasing us with a cheeky glimpse of a much bigger gestalt.

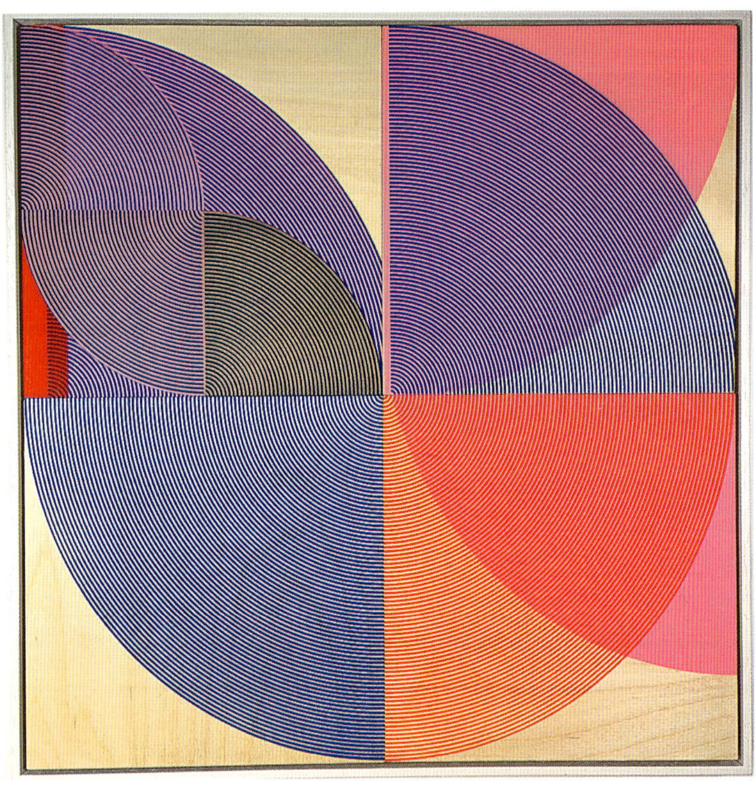

'The main elements to my process are collage, digital textures (geometric patterns developed on the computer) and old test prints. I don't plan my prints that much these days. Instead there is an ongoing process of conversation between sketchbooks, the digital textures I generate, and older prints. Test prints and outtakes tend to get cut up and recycled into collage and sketchbook material.

'I have a growing collection of sketchbooks that are like zines. Each zine is worked on from first page through to last in one big collage session. Through this process of chopping up and re-piecing together, new compositions start to emerge. Once I've spotted some remixed patterns I can turn them into stencils that can be exposed onto screens, and the print process begins again. The notion of "remixing" my own work stems from my interest in DJ and remix culture. Listening to music and hearing the links between patterns and phases all feed into the work I make. In some way I'm trying to capture these sounds and moods in my prints – snapshots of ephemeral moments.'

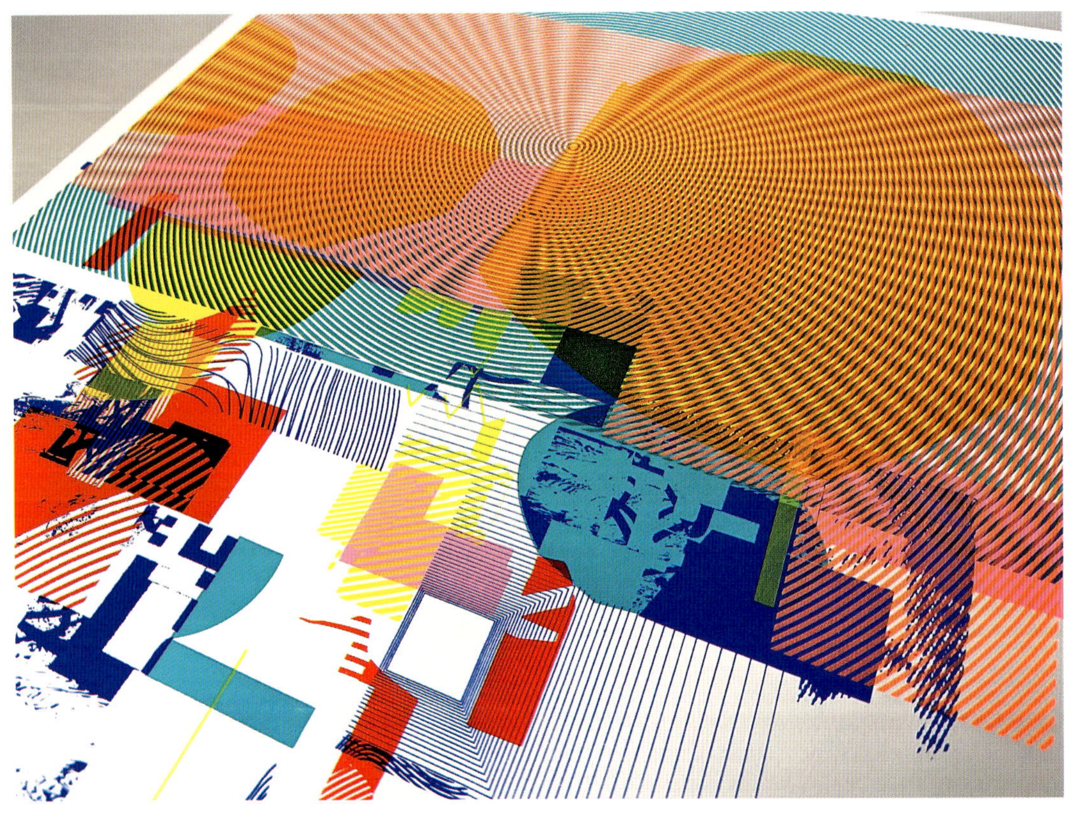

Above left: Iain Perry, *Fractiles*. Screen print

Left: Iain Perry, *Watch the Ride*. Screen print

BRIGHT COLOURS

Bright colours are a good way of creating shape and movement. Optical effects like those seen in Iain Perry's works are made when complementary and contrasting colours are juxtaposed. In this sample I have layered paint, paper, textile and stitch to create a simple block design collage. The waxy and textured surface is the result of adding luminous crayon to the top layer.

Below: Anne Kelly, *Fluoro Colour Collage*. Paper and stitch

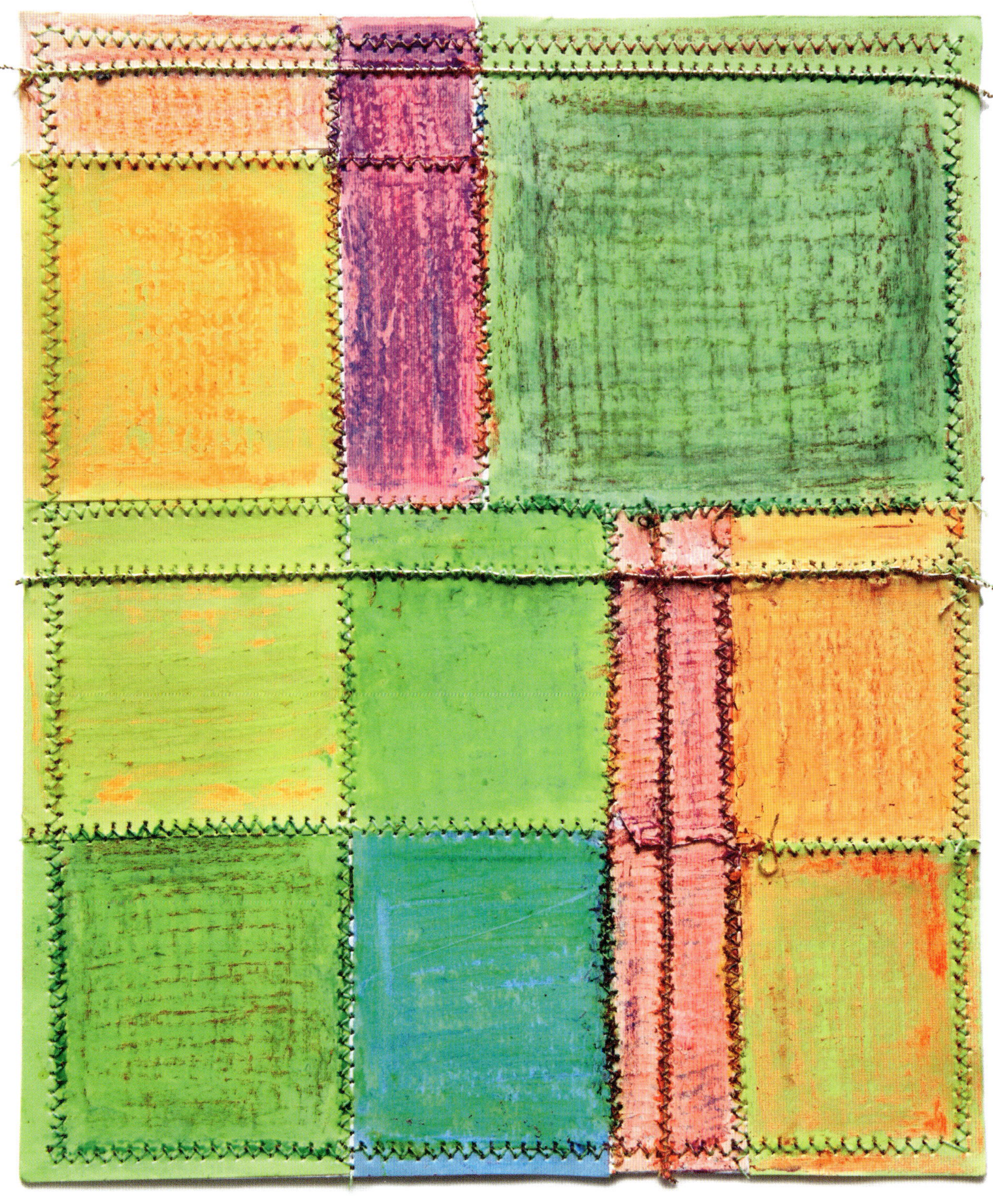

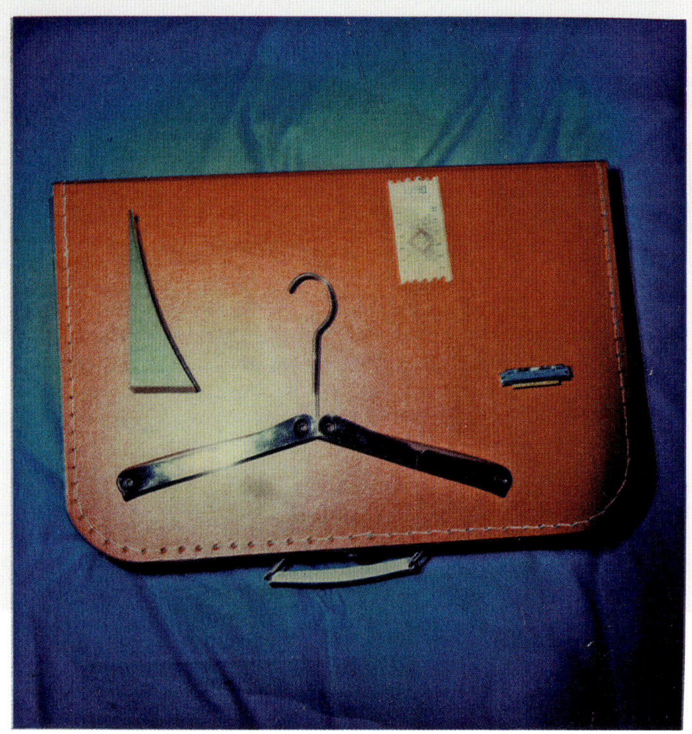

'Being creative is not so much the desire to do something as the listening to that which wants to be done: the dictation of the materials.'

Anni Albers (1899–1994)

CONCLUSION

In this book, more than in any other that I have written or collaborated on, the use of materials has been crucial. It has defined the work and its meaning. I found it refreshing and educative to learn from the contributors' varied approaches to their work. It also belies the commonly held view that somehow abstract work is 'easier' than figurative work. It has been wonderful to see the progression and connections between the two genres.

I am grateful to the contributors for accompanying me on this journey and for sharing their thoughts and practices. I enjoyed seeing how their ideas translated into projects and completed pieces. It is also inspiring to see the work of a previous generation, who were pioneers in abstraction.

I hope that this volume inspires its readers to get into the studio or workspace and try some experiments in abstraction, using the suggested guidelines and workshops peppered throughout the book. Sampling is an encouraging and stabilising way to work, giving you a firm foundation to create new pieces as you go. Enjoy the making, stitching and playing!

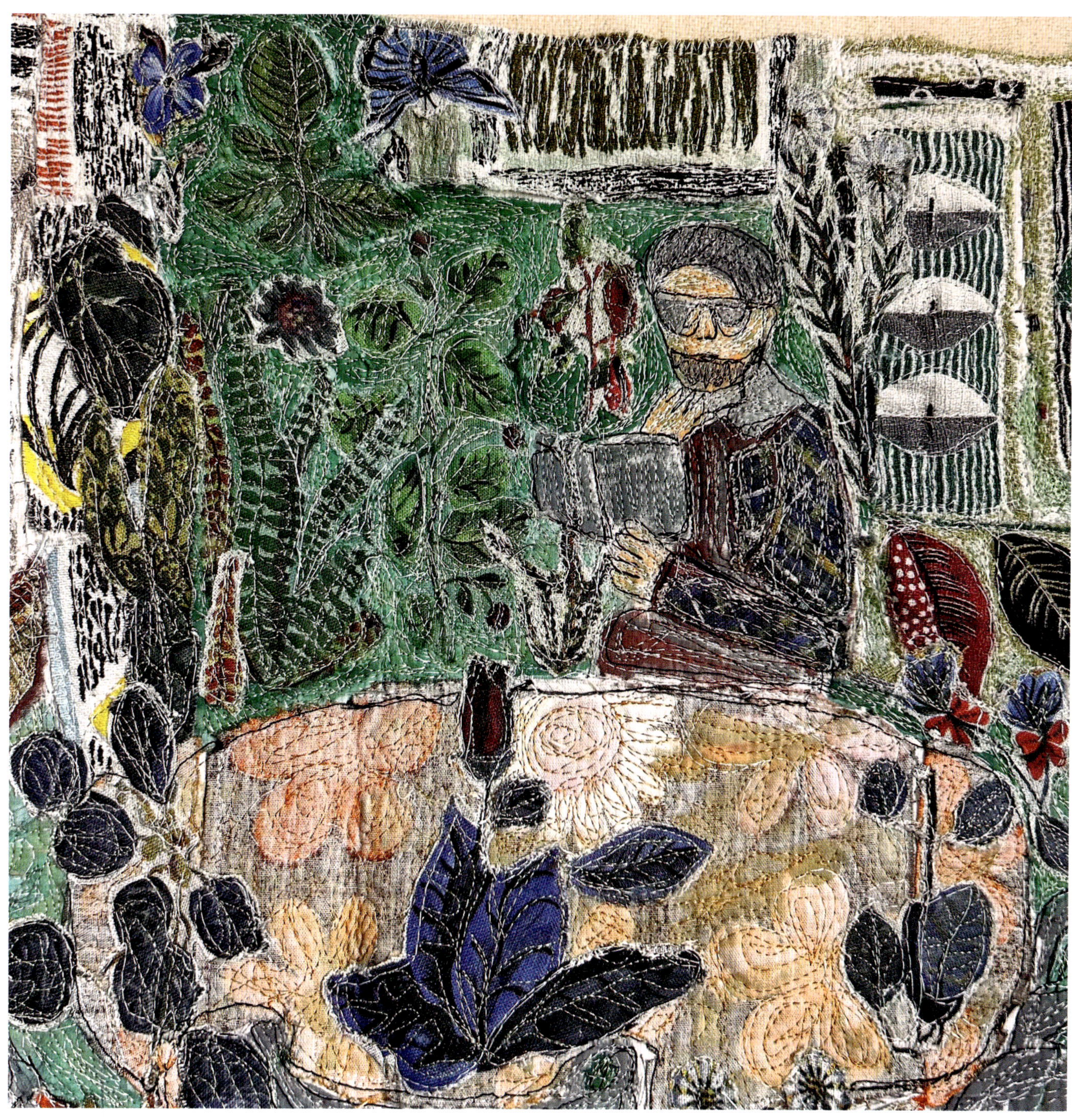

Opposite: Anne Kelly, *Still Life, London*, 1983. Polaroid photograph

Above: Anne Kelly, *Paul in the Garden* (detail). Mixed media textile

FEATURED ARTISTS AND CONTRIBUTORS

Annely Juda Fine Art	www.annelyjudafineart.co.uk
Tara Axford	www.taraaxford.com
Louise Baldwin	www.62group.org.uk/artist/louise-baldwin
Helen Banzhaf	www.helenbanzhaf.co.uk
Jan Beaney	www.62group.org.uk/artist/jan-beaney www.doubletrouble-ent.com
Dail Behennah	www.dailbehennah.com
Eve Campbell	www.evecampbell.co.uk
Lorna Crane	www.lornacrane.com
Jessie Cutts	www.cuttsandsons.com
Jo Elbourne	www.jorobynelbourne.com
Isabel Fletcher	www.isabelfletcher.com
Ruth Fox	Instagram @ruth_fox_stitch
Heather Gratton	Instagram @gratton.heather
Michelle House	www.michellehouse.co.uk
Dalia James	www.daliajames.com
Liske Johnson	Instagram @liske_textileartist
Tim Johnson	www.browngrotta.com/artists/tim-johnson
Jean Littlejohn	www.62group.org.uk/artist/jean-littlejohn www.doubletrouble-ent.com
Louisa Loakes	www.louisaloakes.com
London Transport Museum	www.ltmuseum.co.uk
Alysn Midgelow-Marsden	www.alysnmidgelowmarsden.com
Morii Designs	www.moriidesigns.com
Lesley Patterson-Marx	www.lesleypattersonmarx.com
Iain Perry	www.jealousgallery.com/collections/iain-perry
Margo Selby	www.margoselby.com
Wendy Shaw	Instagram @tickingstripes_sales
Sarah Z. Short	www.sarahzshort.com
Maxine Sutton	www.maxinesutton.com
Sarah Symes	www.sarahsymes.com
Sarah West	www.photoworkx.nl/sarah-west
Deborah White	www.deborahwhite-art.com
Erin Wilson	www.erinwilsonquilts.com

Opposite: Anne Kelly, *Sketchbook Page*. Mixed media

FURTHER RESEARCH

The Amelia	www.theamelia.co.uk
Crafts Council	www.craftscouncil.org.uk
Embroiderers' Guild	www.embroiderersguild.com
Embroidery Magazine	www.embroiderymagazine.co.uk
European Textile Network	www.etn-net.org
London Transport Museum	www.ltmuseum.co.uk
The Quilters' Guild	www.quiltersguild.org.uk
Textile Society of America	www.textilesocietyofamerica.org
Timeless Textiles	www.timelesstextiles.com.au
The Women's Library LSE	www.lse.ac.uk/library/collection-highlights/the-womens-library

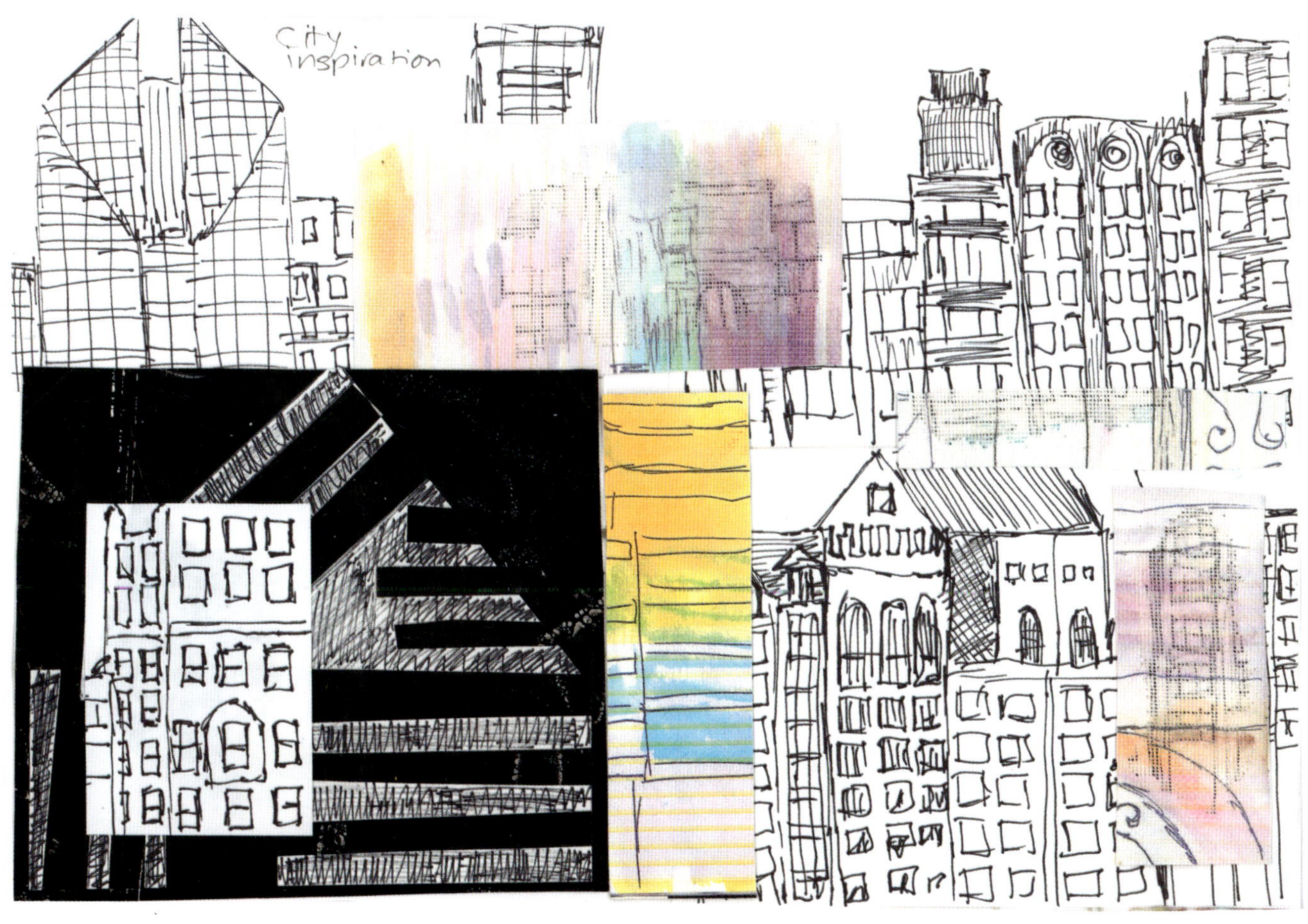

FURTHER READING

Briscoe, Susan, *The Book of Boro*, David and Charles, 2020

Ichi.co, *Small Loom Weaving*, Tuttle, 2022

Kaul, Ekta, *Kantha – Sustainable Textiles and Mindful Making*, Herbert Press, 2024

Kelly, Anne, *Textile Folk Art*, Batsford Books, 2018

Kelly, Anne, *Textile Nature*, Batsford Books, 2016

Kelly, Anne, *Textile Portraits*, Batsford Books, 2023

Kelly, Anne, *Textile Travels*, Batsford Books, 2020

Logan, Jason, *Make Ink – A Forager's Guide to Natural Inkmaking*, Abrams, 2018

Rayner, Geoffrey, *Artists' Textiles in Britain*, ACC Art Books, 1999

Umland, Anne, *Sophie Taeuber-Arp Head*, MoMa Publications, 2019

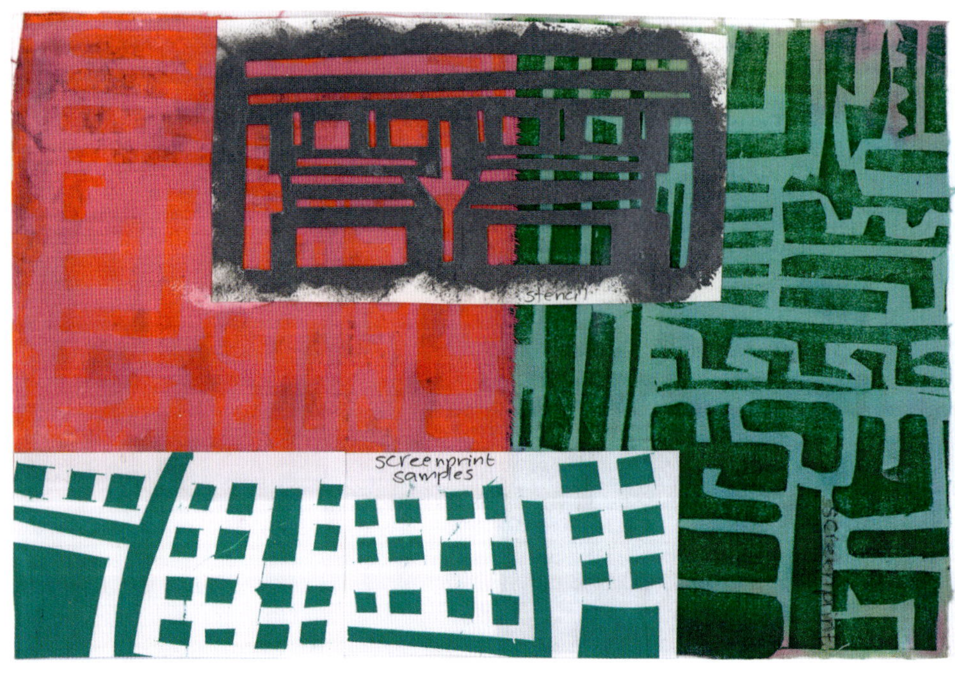

Left: Anne Kelly, *Sketchbook Page*. Mixed media

SUPPLIERS (UK)

Bernina UK
91 Goswell Road
London EC1V 7EX
Tel: 020 7549 7849
info@bernina.co.uk

George Weil
Old Portsmouth Road
Peasmarsh
Guildford GU3 1LZ
Tel: 01483 565800
www.georgeweil.com

Jackson's Art Supplies
1 Farleigh Place
London N16 7SX
Tel: 020 7254 0077
www.jacksonsart.com

Loop London
15 Camden Passage
London N1 8EA
Tel: 020 7288 1160
www.loopknitting.com

Seawhite
Avalon Court
Star Road Trading Estate
Partridge Green RH13 8RY
Tel: 01403 711633
www.seawhite.co.uk

Shepherds Art Supplies
30 Gillingham Street
London SW1V 1HU
Tel: 020 7233 9999
www.store.bookbinding.co.uk/store/

ART TEXTILE COURSES

UK

Bath Textile Summer School
www.bathtextilesummerschool.co.uk

Cowslip Workshops
www.cowslipworkshops.co.uk

Hope & Elvis
www.hopeandelvis.com

West Dean College of Arts and Conservation
www.westdean.org.uk

USA and Canada

French General
www.frenchgeneral.com/collections/workshops

Maiwa School of Textiles
www.maiwa.teachable.com

Australia

Fibre Arts Take two
www.fibreartstaketwo.com

Grampians Arts
www.grampianarts.com.au

INDEX

ACKNOWLEDGEMENTS

My thanks to the artists, makers and contributors named in the text. Their details are in the Featured Artists and Contributors section on page 122.

Special thanks are due to Frida Green and Bella MacConnol at Batsford for their support of and work on this project. A big thank you to Rachel Whiting for her wonderful photography.

To my granddaughter, Sophie Yun, and her family for their love and support.
For my family, especially Paul, and our artistic adventures together.

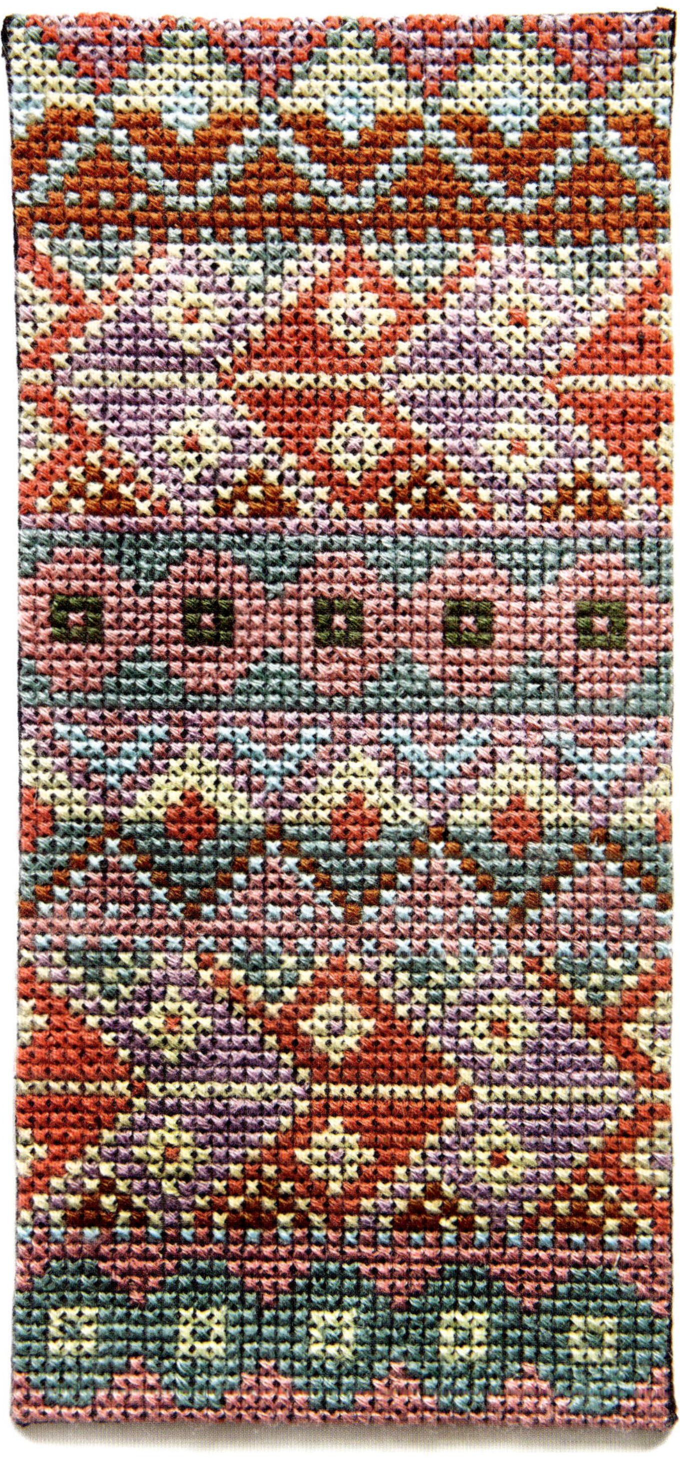

Photo credits

Book photography: Rachel Whiting
Additional photography: Anne Kelly
Enid Marx image (page 45): ©TfL from the London Transport Museum collection
Dail Behennah images (page 51): Jo Hounsome Photography
Dalia James images (pages 76, 77): James Champion
Lorna Crane images (pages 108, 109, 111): Janet Tavener

Right: Ruth Fox, *Cross Stitch*. Embroidery thread on binca